S0-AQO-343

THE COMPLETE MINERAL ENCYCLOPEDIA

THE COMPLETE MINERAL ENCYCLOPEDIA

GRAMERCY BOOKS
New York

Explanation of the abbreviations used in the book

C – color
S – streak
L – luster
D – diaphaneity
DE – density
H – hardness
CL – cleavage
F – fracture
M – morphology
LU – luminescence
R – remark
X – crystal
XX – crystals

In the figure captions the bigger dimension is always mentioned

Scale of the frequency of the occurrence

● - very rare
●● - rare
●●● - uncommon
●●●● - abundant
●●●●● - common

The Complete Mineral Encyclopedia

This 2003 edition published by Gramercy Books, an imprint of Random House Value Publishing, a division of Random House, Inc., New York, by arrangement with Rebo International b.v., Lisse, The Netherlands.

Printed in Slovenia

Random House

New York • Toronto • London • Sydney • Auckland

www.randomhouse.com

A catalog record for this title is available from the Library of Congress.

ISBN 0-517-22169-1

9 8 7 6 5 4 3 2 1

Contents

Introduction

The ever increasing number of publications about minerals reflects a growing interest in nature. Howerer, most of those publications only deal with a few dozen of the most common minerals or gemstones. This book fills this gap by also featuring less common and rare minerals. The authors here describe over 600 mineral species and varieties, illustrated with about 750 color photographs. In choosing illustrations of particular minerals, aesthetic criteria such as size of crystal and color played a role, in addition to their importance and distribution in nature.

This book includes some rare minerals, known only from one locality, because they form very attractive crystals or aggregates. There are minerals known to humankind since prehistoric times, such as quartz and gold, but also minerals first discovered quite recently like rossmanite. The photographs show not only well-formed and colorful crystals but many aggregates, which are more common in nature.

The minerals in this book are listed according to the mineralogical system of Hugo Strunz, in his book *Mineralogische Tabellen* (1978). The chemical formulae of individual minerals follow the form of *Glossary of Mineral Species* (1995) by M. Fleischer and J.A. Mandarino. The information is complemented in both cases with the latest knowledge from scientific literature, such as new nomenclature of amphiboles, micas and zeolites.

The mineral descriptions cover the basic physical and chemical data, including chemical formulas and crystal systems. The data provided corresponds mainly to the end-members. The less common valence of the chemical elements is marked in the chemical formula (Fe^{3+}, Mn^{3+}, As^{3+}, Mn^{4+}, Pb^{4+}). Where an element features both valences in the mineral, they are both marked (e.g. ilvaite, braunite).

The origins of the individual minerals are described in detail. We chose for a relatively simplified scheme because the normal complexity cannot be described in detail. Minerals can be distinguished as either primary (resulting directly from a solidifying of magma, crystallizing of an aqueous solution, or metamorphism – re-crystallization in a solid state) and secondary (resulting from alteration of the original mineral, e.g. during its oxidation or reduction under low temperature and pressure close to the surface of the Earth). Primary minerals are divided into the following groups: 1. magmatic, when a mineral crystallizes directly from a melt (it includes magmatic and effusive igneous rocks, including granitic and alkaline syenite pegmatites and meteorites); 2. sedimentary, when a mineral crystallizes during a process of diagenesis or from hydrous solutions under normal temperature (clastic, organic and chemical sedimentary rocks); 3. metamorphic, when a mineral crystallizes during metamorphic processes in a solid state at a wide range of temperature and pressure (it includes regionally and contact metamorphosed rocks and skarns); 4. hydrothermal, when minerals crystallize from aqueous solutions and fluids under high to low temperatures (it includes ore and the Alpine-type veins, cavities in volcanic rocks, minerals and rocks, hydrothermally altered under high temperature, e.g. greisens).

Secondary minerals are divided into the following groups: 1. oxidation, when minerals result from the oxidation (weathering) of the primary minerals in the oxidation zone of ore deposits and other rocks (it includes the origin of malachite and azurite during the chalcopyrite oxidation, also the origin of secondary phosphates in granitic pegmatites during the oxidation of primary phosphates); 2. cementation, when minerals result from the reduction of the primary minerals (the origin of native copper and native silver under the reduction conditions in the cementation zone of ore deposits). This classification is very much simplified, because in many cases we cannot readily determine a specific origin of a particular mineral. This relates to minerals that crystallize under conditions which approximately represent a transition between separate phases of the origin, such as the magmatic or hydrothermal origins of elbaite in the pegmatite cavities; the metamorphic or hydrothermal origins of grossular in skarns; the magmatic or metamorphic origins of cordierite in migmatites; and the hydrothermal or secondary origins of some phosphates in granitic pegmatites etc.

With the localities of individual minerals we have tried to list the most important worldwide localities regardless of their recent production, but we have also included recent discoveries, since these may produce important mineral specimens. Where a mineral has an important use, this is listed at the end of mineral description.

We would like to acknowledge all those who contributed in any way to the production of this book, particularly the private collectors and institutions which loaned minerals for photography. We hope those fascinated in the world of minerals and in nature will find this book a fascinating source of information.

This book is dedicated to the memory of Dr. Jaroslav Senek, who was of extraordinary influence to several generations of Czech and Slovak mineralogists and mineral collectors, for his enthusiasm for mineralogy and his attitude on life.

Marcasite, 100 mm, Misburg, Germany

1. Elements

Copper

Cu

CUBIC ● ● ● ● ●

Properties: C – light pink to copper-red, it darkens and covers green to black in air; S – red; L – metallic; D – opaque; DE – 8.9; H – 2.5 – 3; CL – none; F – hackly; M – cubic crystals and its combinations, dendritic aggregates, sheets, slabs, massive.
Origin and occurrence: Primary hydrothermal copper; is mainly related to basic igneous rocks; it is also common as a product of supergene cementation. It is associated with cuprite, malachite, azurite, silver, chalkocite, bornite, and other minerals. The largest accumulations of primary copper are in the Keweenaw Peninsular, Lake Superior, USA, the largest being 15 x 7 x 3m (approx. 50 x 23 x 10ft) and weighing 420 tons. Fine crystals up to 50 mm (approx. 2in) also occur there, as do calcite crystalwith copper inclusions. Superb supergene coppercrystals come from many localities like Tsumeb, Namibia and Chessy, France. Crystals up to 140mm (5½ in) long occurred in the Ray mine and in Bisbee, Arizona, USA. Very fine spinel-law twins up to 5 cm in size and dendritic aggregates come from Mednorudnyansk, Ural mountains, Russia; crystals up to 30 mm (13/16 in) were found in Dzhezkazgan, Kazakhstan. Fine specimens of copper, associated with cuprite, azurite and malachite occurred in Rudabánya, Hungary. *Application:* electronics, electrical engineering, ingredient in gold alloys.

Copper, 42 mm, Cornwall, UK

Silver, 52 mm, Freiberg, Germany
Copper, 120 mm, Keweenaw Peninsula, U.S.A.

Silver
Ag

CUBIC ● ● ●

Properties: C – silver-white, tarnishes gray to black; S – silver-white; L – metallic; D – opaque; DE – 10,5; H – 2.5 – 3; CL – none; F – hackly; M – cubic crystals, dendritic aggregates, wires, leaves, massive.

Origin and occurrence: Hydrothermal in ore veins and also of secondary cementation origin in association with acanthite, stephanite, proustite, pyrargyrite, copper and many other minerals. The best specimens of crystallized and wire silver come from Kongsberg, Norway, where wires up to 400mm (16 in) long and crystals up to 40 mm ($1^{9}/_{16}$ in) in size have been found. Beautiful specimens of wire silver with wires over 100 mm (4 in) long are known from Freiberg, Schneeberg and St. Andreasberg, Germany. Wires several cm long were also found in Príbram and Jáchymov, Czech Republic. Dendritic aggregates from Batopilas, Chihuahua, Mexico, reached up to 150 mm (6 in). Crystals and aggregates of silver, grown together with copper are genetically unique in the basalt cavities in the Keweenaw Peninsula near Lake Superior, Michigan, USA. Wires, up to 100 mm (4 in) long, come also from the San Genaro Mine in Huancavelica and Uchucchaqua, Peru. New finds of silver wires, up to 150 mm (6 in) long, have been made in Dzhezkazgan, Kazakhstan.

Application: photographic industry, jewelry, electronics.

Silver, 55 mm, Schwarzwald, Germany

Gold, 48 mm, El Dorado, California, U.S.A.

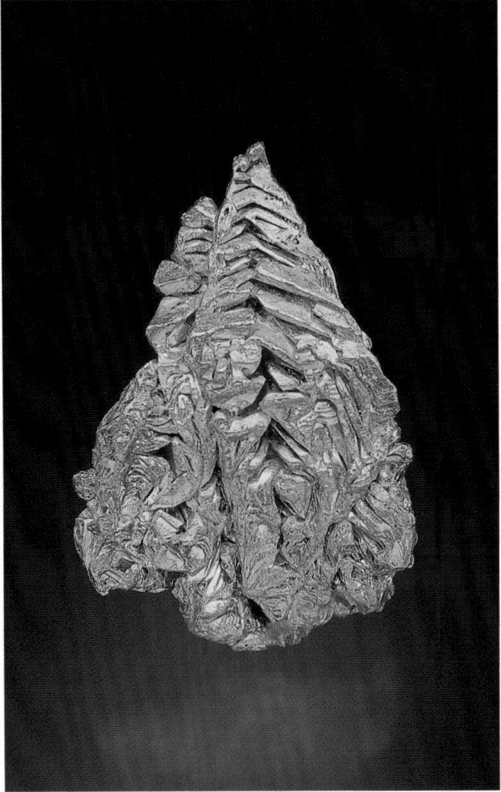

Gold, 68 mm, Eagle's Nest Mine, California, U.S.A.

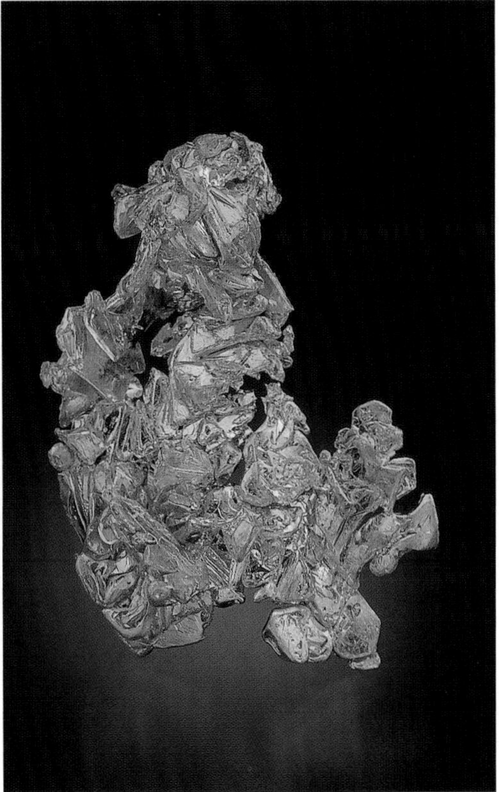

Gold

Au

CUBIC ● ●

Properties: C – gold-yellow; S – yellow; L – metallic; D – opaque; DE – 19.3; H – 2.5-3; CL – none; F – hackly; M – octahedral and cubic crystals, skeletal and dendritic aggregates, leaves, nuggets.
Origin and occurrence: Primary hydrothermal in ore veins, also in contact with metamorphic deposits and pegmatites. Placer deposits are secondary. It occurs with pyrite, arsenopyrite, quartz, sylvanite, calaverite, krennerite and other minerals. Beautiful leaves and crystals of gold found in many localities in California, USA (Colorado Quartz mine, Negro Hill and others). Fine leaf gold comes from Rosia Montana, Romania. The best crystals, skeletal octahedra, up to 50 mm (2 in) have been found in alluvial sediments near Gran Sabana, Roraima Shield, Venezuela. Gold wires up to 110 mm (4 5/16 in) long were rarely in Ground Hog mine, Gilman, Colorado, USA. The largest known sheets of crystallized gold occurred in the Jamestown mine, California, USA, where a cavity, which yielded 49 kg (108 lb) of golden leaves, was discovered in 1992. The largest measures about 300 mm (11 13/16 in) and has about 25.79 kg (56 lb 13 oz) of gold on it. Typical aggregates of fine gold wires come from Farncomb Hill near Breckenridge, Colorado, USA. Fine crystals were also found in Berezovsk, Ural mountains and in the Lena River basin, Siberia, Russia. Fine scales and larger nuggets from placer deposits were found in Klondike, Alaska; Tuolumne County, California, USA and in Ballarat, Victoria, Australia. Fine dendritic aggregates occurred in the Hope's Nose, Devon, UK. A unique find of leaves up to 100 mm (4 in) was made in Krepice near Vodnany, Czech Republic.
Application: practically the only source of gold as a metal; used in jewelry, electronics and medicine.

Mercury

Hg

TRIGONAL ● ●

Properties: C – tin white; L – metallic to adamantine; D – opaque; DE – 13.6; M – liquid at temperatures above -39°C (-38.2°F); R – very poisonous fumes.
Origin and occurrence: Hydrothermal in low-temperature ore deposits, also connected with hot springs. It is associated with cinnabar, calomel and other Hg minerals. It occurred in Almaden, Spain;

Mercury, 30 mm, Socrates Mine, California, U.S.A.

Idria and Avala, Serbia; New Almaden and New Idria, California; Terlingua, Texas, USA; Dedova hora, Czech Republic; and Rudnany, Slovakia as droplets and liquid cavity fillings.
Application: chemical industry, measuring instruments, metallurgy.

Moschellandsbergite
$Ag_2 Hg_3$

CUBIC ● ●

Properties: C – silver-white; S – silver-white; L – metallic; D – opaque; DE – 13.5; H – 3.5; CL – good; F – conchoidal; M – dodecahedral crystals and their combinations, granular, massive.
Origin and occurrence: Hydrothermal in low-temperature deposits, associated with cinnabar, tetrahedrite, pyrite and other minerals. Crystals, several mm long, were found in Moschellandsberg, Germany. They are also known from Sala, Sweden; Les Chalanches, France and Brezina, Czech Republic.

Moschellandsbergite, 6 mm grain, Moschellandsberg

Lead
Pb

CUBIC ● ●

Properties: C – gray-white, tarnishes to lead-gray and gets dull; S – lead-gray; L – metallic; D – opaque; DE – 11.3; H – 1.5; CL – none; M – octahedral and cubic crystals, massive.
Origin and occurrence: Hydrothermal, also sedimentary (authigenic), associated with willemite and other minerals. The best specimens with crystals up to 40 mm (1 9/16 in) in size, come from Langban, crystallized also from Pajsberg, Sweden. Octahedra, up to 10 mm (3/8 in), are described from El Dorado, Gran Sabana, Venezuela. It also occurs in Franklin, New Jersey, USA and Jalpa, Zacatecas, Mexico.

Iron
Fe

CUBIC ● ●

Properties: C – steel-gray to black; S – gray; L – metallic; D – opaque; DE – 7.9; H – 4; CL – perfect; F – hackly; M – crystals, granular, massive.
Origin and occurrence: Terrestric iron occurs mainly in basic rocks, but it is also known from carbonate sediments and the petrified wood. The most famous locality is Blaafjeld near Uivfaq on Disko Island, Greenland, where masses up to 20 tons were found.

Lead, 53 mm, Langban, Sweden

Iron, 80 mm, Bühl, Germany

Chunks, weighing over 10 kg (22 lb) come from Bühl near Kassel, Germany. Impregnations of iron in dolerite occur in the Khuntukun massif and masses up to 80 kg (176 lb) are known from Ozernaya Mt., Siberia, Russia.

Platinum

Pt

CUBIC ● ●

Properties: C – steel-gray to dark gray; S – steel-gray to silver-white; L – metallic; D – opaque; DE – 21.5; H – 4-4.5; CL – none; F – hackly; M – cubic crystals, nuggets, grains and scales.

Platinum, 8 mm, Konder, Russia

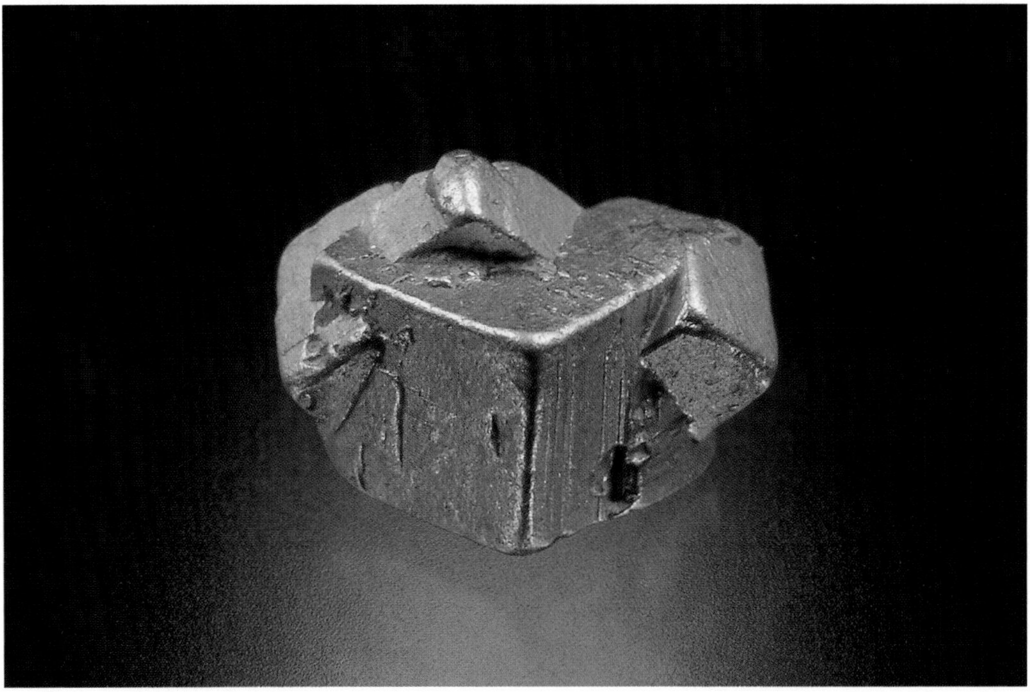

Origin and occurrence: Platinum occurs in magmatic segregations, together with chromite, olivine and magnetite in ultrabasic rocks; secondary in placers. The best crystals up to 15 mm ($^{19}/_{32}$ in) come from Konder in Khabarovsk Region, small crystals, but mainly nuggets, weighing up to 11.5 kg (25 lb 5 oz) found in the Tura River basin near Turiinsk, Ural Mountains. Primary platinum is known from deposits in the vicinity of Nizhniy Tagil, Ural Mountains, Russia; from the Onverwacht mine, Bushveld, South Africa and Sudbury, Ontario, Canada. Fine smaller nuggets, weighing up to 75 g (165 lb) were found in the Trinity River sediments in California, USA and the Chocó River sediments in Columbia.
Application: chemical industry, catalytic convertors, rocket industry.

Arsenic, 45 mm, Alden Island, Canada

Arsenic
As

TRIGONAL ● ● ●

Properties: C – tin-white, tarnishes quickly to black; SB tin-white; L – metallic; D – opaque; DE – 5.8; H – 3.5; CL – perfect; F – uneven; M – rhombohedral crystals, botryoidal aggregates, granular, massive.
Origin and occurrence: Mainly hydrothermal, together with other As minerals. It forms massive veins, up to 200 mm ($7^7/_8$ in) thick, with botryoidal surface in Jáchymov, Czech Republic and in Freiberg, Germany. In Akatani, Japan, spherical aggregates consisting of small crystals were found. It is also known from Sacarímb, Romania. Botryoidal aggregates of arsenic with leaf gold were found in the Royal Oak mine, Coromandel, New Zealand. Crystals of metamorphic origin come from Sterling Hill, New Jersey, USA.

Stibarsen, 58 mm, Atlin, Canada

Antimony, 37 mm, New Brunswick, Canada

Stibarsen
SbAs

TRIGONAL ● ● ●

Properties: C – tin-white to gray, tarnishes black; S – gray; L – metallic, sometimes dull; D – opaque; DE – 6.3; H – 3-4; CL – perfect; M – indistinct crystals, botryoidal aggregates.
Origin and occurrence: In pegmatites with antimony, stibiotantalite and microlite; hydrothermal in ore veins withpyrargyrite, proustite, pyrostilpnite and dyscrasite. Beautiful botryoidal aggregates of stibarsen (previously labeled as allemontite), up to 100 mm (approx. 4 in) in size, come from Príbram and Trebsko, Czech Republic. Botryoidal aggregates up to 80 mm ($3^1/_8$ in) and imperfect crystals were found in quartz veins in Atlin, British Columbia, Canada. Fine specimens occurred in a Li-bearing pegmatite near Varuträsk, Sweden.

Antimony
Sb

TRIGONAL ● ● ●

Properties: C – tin-white; S – gray; L – metallic; D – opaque; DE – 6.7; H – 3-3.5; CL – perfect; F – uneven; M – rhombohedral crystals, botryoidal aggregates, massive.
Origin and occurrence: Hydrothermal in ore veins with silver, stibnite, stibarsen, sphalerite and other minerals; also in pegmatites. As veinlets in a pegmatite in Varuträsk, Sweden. Cleavable plates, up to 50 mm (2 in) known from Törnäva, Finland.

Massive aggregates up to 200 mm (7 7/8 in) come from Príbram, Czech Republic. Rhombohedral crystals up to 10 mm (3/8 in) across and accumulations up to 300 mm (11 13/16 in) in size described from Lake George, New Brunswick, Canada.

Bismuth

Bi

TRIGONAL ● ● ●

Properties: C – silver-white, tarnishes pink; S – silver-white; L – metallic; D – opaque; DE – 9.8; H – 2-2.5; CL – perfect; M – rhombohedral crystals, granular, massive.
Origin and occurrence: It is found in pegmatites, greisens and hydrothermal in ore veins together with chalkopyrite, arsenopyrite, löllingite, nickeline, breithauptite and many other minerals. Common in pegmatites in Anjanabonoina, Madagascar. Very fine crystals, up to 20 mm (25/32 in) known from Schlema and Hartenstein, Germany. Skeletal aggregates, overgrown with other arsenides, occurred in Jáchymov, Czech Republic. Masses weighing several kg found in Bolivia (Tasna, Velaque) and Australia (Kingsgate, New South Wales). Cleavable masses up to 12 cm (4 11/16in) described from Cobalt and Gowganda, Ontario, Canada.
Application: Bi ore.

Arsenolamprite

As

ORTHORHOMBIC ●

Properties: C – gray-white, it covers with a black coating; S – black, L – metallic to adamantine; D – opaque; DE – 5.6; H – 2; CL – perfect; M – acicular crystals, tabular and fan-shaped aggregates, massive.
Origin and occurrence: Hydrothermal in ore veins associated with arsenic, bismuth, silver and other minerals. Its crystals and veinlets were found with Cu arsenides in Cerny dul, Czech Republic. Occurs also in Jachymoc, Czech Republic and Marienberg Germany. Found recently in Cavnic, Romania.

Bismuth, 20 mm, Cínovec, Czech Republic

Arsenolamprite, 70 mm, Jáchymov, Czech Republic

Graphite

C

HEXAGONAL ● ● ● ● ●

Properties: C – black to steel-gray; S – black to steel-gray; L – metallic, dull, earthy; D – opaque; DE – 2.3; H – 1-2; CL – perfect; M – hexagonal tabular crystals, massive.
Origin and occurrence: Metamorphic, from metamorphism of a sedimentarymaterial with C contents: also primary magmatic.Associated with many materials, stabile underconditions of the graphite origin. Crystals several cm in size known from Nordre Stromfjord, Greenland. Crystals were also found in Sterling Hill, New Jersey and Crestmore, California, USA Foliated aggregates are found in Sri Lanka (Radegara, Galle region). Accumulations in Buckingham and Grenville, Quebec, Canada are industrially important. Also common in Shunga deposit in Karelia, Russia; in Cesky Krumlov, Netolice and Blizná, Czech Republic.
Application: metallurgy, nuclear industry, production of lubricants.

Graphite, 130 mm, Krichim, Bulgaria

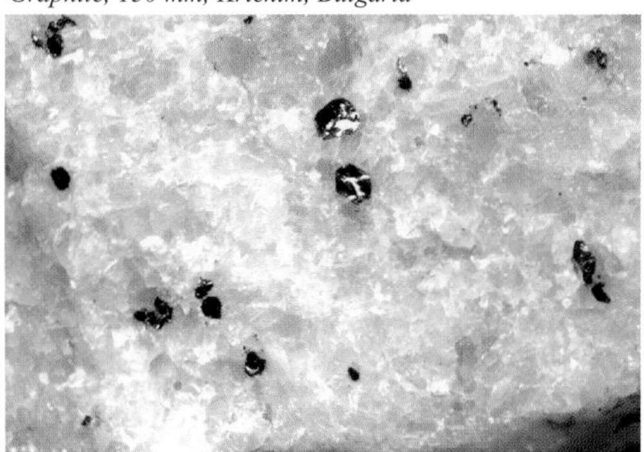

Diamond

C

CUBIC ● ●

Varieties: bort (opaque technical diamonds), balas (dark colored, spherical radial aggregates), carbonado (brown-black to black massive aggregates, up to egg sized).

Properties: C – colorless, yellow, brown, white, pink, black, red, blue, green; S – white; L – adamantine; D – transparent to translucent; DE – 3.5; H – 10; CL – perfect; F – conchoidal; M – octahedral and cubic crystals; LU – sometimes fluorescent, sometimes phosphorescent.

Origin and occurrence: Primary magmatic occurrences are limited to kimberlite pipes, secondary occurrences to placers. Large primary deposits are known from South Africa (Premier mine, Kimberley) and Yakutia, Russia (pipes in the vicinity of Mirnyi). Primary and secondary occurrences of diamonds are located in lamproites and placers near Argyle, Western Australia. Most historical diamonds from India (Golconda), Brazil (Diamantina, Minas Gerais), Congo, Angola and Namibia were found in placer deposits. Diamonds of industrial grade always prevail over the gem-quality stones. The largest gem-grade diamond ever found, the Cullinan, weighing 3106 carats, comes from the Premier mine in Kimberley, South Africa. It yielded gem rough for 104 faceted stones, the heaviest of which weighs 531 carats. The largest faceted diamond known, called Golden Jubilee, was found in the same place in 1986. It weighed 755 carats before cutting and as a finished stone it weighs 545.65 carats. The dark blue Hope (44 carats) and the green Dresden (76 carats) probably came from India. The unique red diamond, weighing 5 carats, which is at the Smithsonian Institution, Washington, DC, USA, is of unknown origin.

Application: the most popular gemstone, bort and carbonado varieties are used as abrasives.

Sulfur

S

ORTHORHOMBIC ● ● ● ●

Properties: C – sulfur-yellow, yellow-brown, greenish, reddish to yellowish-gray; S – white; L – resinous to greasy; D – transparent to translucent; DE – 2.1; H – 1.5-2.5; CL – imperfect; F – conchoidal to uneven; M – dipyramidal, disphenoidic and thick tabular crystals, botryoidal and stalactitic aggregates.

Origin and occurrence: Hydrothermal product of fumaroles, product of an activity of microorganisms, disintegration of sulfides and acidic chemical reactions; associated with gypsum, anhydrite, aragonite, calcite,

Diamond, 15 mm x, Mirnyi, Russia

Sulphur, 66 mm, Sicily, Italy

celestite and halite. The world's best crystals come from many localities near Girgenti, Sicily, Italy (Caltanisseta, Cianciana) where they reached up to 12 cm ($4\frac{11}{16}$in) in size. Fine crystals are also known from Tarnobrzeg, Poland and Yavokskoye near Lvov, Ukraine. As a product of solfataras it occurs in many volcanically active places, like Solfatara near Pozzuoli, Italy or in sulfur lava near Shiretoku, Japan. Sulfur layers, up to 30 m (100 ft) thick, associated with salt diapirs, are located near Charles Lake, Louisiana, USA. It originates during intensive oxidation/reduction reactions of pyrite in Rio Tinto, Spain and in Kostajnik, Serbia.
Application: chemical, paper-making, rubber and leather-making industries, agriculture.

Selenium, 110 mm, Kladno, Czech Republic

Selenium
Se

TRIGONAL ● ●

Properties: C – gray to red-gray; S – red; L – metallic, D – opaque to translucent; DE – 4.8; H – 2; CL – good; M – acicular crystals, droplets of vitreous surface, felt-like aggregates.
Origin and occurrence: Secondary, resulting from alteration, fumaroles and from burning coal dumps, with sulfur, sal ammoniac and other sulfates. Also from oxidation of organic compounds in U- and V-bearing deposits of the Colorado Plateau type, associated with pyrite, zippeite and other minerals. Red needles up to 20 mm ($\frac{25}{32}$ in) long come from the United Verde mine, Jerome, Arizona, USA. Black selenium needles found in burning coal dumps in Kladno and Radvanice, Czech Republic. Occurs with ores of U and V along sandstone fissures in the Peanut Mine, Bull Canyon, Colorado, USA. Occurred through volcanic activity in Vulcano, Lipari Islands, Italy.

Tellurium
Te

TRIGONAL ● ●

Properties: C – tin-white; S – gray; L – metallic; D – opaque; DE – 6.2; H – 2-2.5; CL – perfect; M – prismatic and acicular crystals, granular, massive.
Origin and occurrence: Primary hydrothermal in low-temperature ore deposits; it originates also as secondary through the oxidation/reduction reactions of tellurides. It is associated with gold, sylvanite, altaite, pyrite and other minerals. Crystals up to 30 mm ($1\frac{3}{16}$ in) long are known from Balya, Turkey. Crystals up to 20 mm ($\frac{25}{32}$ in) long occurred in the Au deposits Cripple Creek and Colorado City, USA. Crystals up to 10 mm ($\frac{3}{8}$ in) across come from Kawazu and Suzuki, Japan. Rich cleavage masses and crystals up to 70 mm (2¾ in) long were found in Uzbekistan. Crystallized tellurium is also known from Fata Baii and Baia de Aries, Romania.

Tellurium, 1 mm xx, Zlatna, Romania

2. Sulfides

Algodonit

Cu_6As

HEXAGONAL ● ●

Properties: C – steel-gray to silver-white, it quickly covers with a brown coating on air; S – gray; L – metallic; D – opaque; DE – 8,7; H – 4; CL – none; F – conchoidal; M – crystals, granular, massive.
Origin and occurrence: Hydrothermal, mainly intimately inter-grown with other Cu arsenides. Its largest accumulations are known from Cu deposits in melaphyres in the Keweenaw Peninsula, Lake Superior, Michigan, USA. It is also known from Chile (Algodones mine near Coquimbo, Atacama). Other localities are Talmessi, Iran and Långban, Sweden.

Domeykite

Cu_3As

CUBIC ● ● ●

Properties: C – tin-white to steel-gray, it tarnishes yellow and covers with a brown coating; S – gray; L – metallic; D – opaque; DE – 7,9; H – 3-3,5; CL – none; F – uneven; M – botryoidal aggregates, massive.
Origin and occurrence: Hydrothermal with copper, cuprite, algodonite and silver. Common in masses, weighing several kg, together with algodonite in the Keweenaw Peninsula, Michigan, USA. The largest accumulations are known from Talmessi and Anarak, Iran. It occurs near Copiap and Chañarcillo, Chile. Massive aggregates in cuprite up to 50 mm (2 in) come from Biloves, Czech Republic.

Sphalerite, 56 mm, Picos de Europa, Spain
Algodonite, 60 mm, Keweenaw Peninsula, U.S.A.

Domeykite, 3 mm x, Rudabánya, Hungary

Allargentum

$Ag_{1-x}Sb_x$

HEXAGONAL ●

Properties: C – silver-white; S – gray; L – metallic; D – opaque; DE – 10,1; H – not determined; CL – none; M – small crystals, granular.
Origin and occurrence: Hydrothermal in the silver-bearing ore veins, associated with silver, breithauptite and dyscrasite. Crystals up to 1 mm ($^1/_{32}$ in) known from Hartenstein, Germany. Its inter-growths with silver ores were found in Cobalt, Ontario, Canada. It is also known from Broken Hill, New South Wales, Australia and microscopic in Rejská vein in Kutná Hora, Czech Republic.

Allargentum, 50 mm, Schlema, Germany

Dyscrasite
Ag_3Sb

ORTHORHOMBIC ● ●

Properties: C – silver-white, it tarnishes yellow to black; S – silver-white; L – metallic; D – opaque; DE – 9,7; H – 3,5-4; CL – good; F – uneven; M – pyramidal and prismatic crystals, granular, massive.
Origin and occurrence: Hydrothermal in the ore veins, associated with silver, stibarsen, pyrargyrite, calcite and other minerals. The best specimens come from the silver-bearing veins, cross-cutting the U deposit Háje near Príbram, Czech Republic, where prismatic crystals up to 50 mm (approx. 2 in) long and striated tabular twins were found. They are mostly embedded in stibarsen; all the specimens, appearing in the mineral shows, are etched out of matrix. Deformed crystals of completely different habit are known from St. Andreasberg, Germany. Crystals occurred also in the Consols mine in Broken Hill, New South Wales, Australia.

Chalcocite
Cu_2S

MONOCLINIC ● ● ● ●

Properties: C – lead-gray to black; S – lead-gray to black; L – metallic; D – opaque; DE – 5,8; H – 2,5-3; CL – imperfect; F – conchoidal; M – prismatic to tabular crystals, granular, massive.
Origin and occurrence: Hydrothermal, also sedimentary and metamorphic, mostly secondary, in the oxidation and cementation zones of ore deposits. It occurs together with pyrite, chalcopyrite, covellite, bornite and other minerals. Crystals up to 25 cm (9 7/8 in) found in the M'Sesa mine, Zaire. Beautiful crystals several cm across come from Redruth and St Just, Cornwall, UK. Crystals over 20 mm (25/32 in) across occurred in Bristol, Connecticut and in Butte, Montana. Crystals up to 50 mm (2 in) in size known from the Flambeau mine near Ladysmith, Wisconsin, USA. Shiny cyclic twins of crystals up to 20 mm (25/32 in) found in Dzhezkazgan, Kazakhstan. Massive aggregates are important Cu ore in Rio Tinto, Spain; Bor, Serbia; Bisbee, Arizona, USA; and Tsumeb, Namibia.
Application: important Cu ore.

Djurleite
$Cu_{31}S_{16}$

MONOCLINIC ● ● ●

Properties: C – lead-gray to black; S – lead-gray; L – metallic; D – opaque; DE – 5,8; H – 2,5-3; CL – none; F – conchoidal; M – short prismatic to tabular crystals, granular, massive.
Origin and occurrence: Secondary as a product of the cementation zone in ore deposits. Crystals up to 10 mm (3/8 in) across known from the Botallack mine

Dyscrasite, 75 mm, Príbram, Czech Republic

Chalcocite, 17 mm x, Cornwall, UK

near St. Just, Cornwall, UK. Aggregates of thick acicular crystals up to 30 mm (1 3/16 in) long found in Dzhezkazgan, Kazakhstan. It occurs in massive form in many porphyry copper deposits (Butte, Montana; Bisbee, Arizona, USA); also in Tsumeb, Namibia, together with chalcopyrite, pyrite and other minerals.

Djurleite, 3 mm xx, Dzhezkazgan, Kazakhstan

Berzelianite

Cu_2Se

CUBIC ● ● ●

Properties: C – silver-white, tarnishes black; S – silver shiny; L – metallic; D – opaque; DE – 7,3; H – 2,5; CL – none; F – uneven; M – granular, massive. *Origin and occurrence:* Hydrothermal, together with other selenides in U, Fe and Au deposits. It is the main mineral in the selenide mineralization in Tilkerode, Germany. Grains, up to several tens of cm across, greenish tarnished, occurred together with other selenides in Bukov, Habrí, Petrovice and Predborice, Czech Republic. Similar occurrence is known from near Pinky Fault near Athabasca Lake, Saskatchewan, Canada.

Berzelianite, 60 mm, Bukov, Czech Republic

Bornite, 15 mm xx, Dzhezkazgan, Kazakhstan

Bornite

Cu_5FeS_4

ORTHORHOMBIC ● ● ● ●

Properties: C – copper-red, tarnishes iridescent; S – gray-black; L – metallic; D – opaque; DE – 5,1; H – 3-3,5; CL – imperfect; F – uneven to conchoidal; M – pseudo-cubic octahedral crystals, massive.
Origin and occurrence: Magmatic, hydrothermal, sedimentary, in skarns and pegmatites together with chalcocite, chalcopyrite, pyrite, quartz and other minerals. Fine crystals up to 10 mm ($^{3}/_{8}$ in) across are known from Carn Brea, Cornwall, England, UK. Crystals up to 30 mm ($^{13}/_{16}$ in) come from Likasi, Shaba, Zaire. Beautiful crystals up to 40 mm ($1^{9}/_{16}$ in) across were found recently together with chalcocite in Dzhezkazgan, Kazakhstan. Massive aggregates are common and used as Cu ore in Kipushi, Shaba, Zaire. Fine-grained, sedimentary bornite occurs in Cu-bearing shales in Mansfeld, Germany, where it forms the main ore layer. Crystals, up to 20 mm ($^{25}/_{32}$ in) in size, occurred in the Cole shaft and masses, weighing several thousands of tons, were mined in the Campbell shaft, Bisbee, Arizona, USA. *Application:* Cu ore.

Bornite, 3 mm xx, Dzhezkazgan, Kazakhstan

Umangite

Cu_3Se_2

TETRAGONAL ● ● ●

Properties: C – blue-black with reddish tint, tarnishes purple; S – black; L – metallic; D – opaque; DE – 6,6; H – 3; CL – imperfect; F – uneven to conchoidal; M – granular, massive.
Origin and occurrence: Hydrothermal in ore veins together with other selenides (clausthalite, berzelianite). It is common, associated with berzelianite in Tilkerode, Germany; Sierra de Umango, Argentina and Slavkovice, Czech Republic. Larger accumulations occur in the Martin Lake mine near Athabasca Lake, Canada.

Acanthite
Ag_2S

MONOCLINIC ● ● ●

Properties: C – black; S – black; L – metallic; D – opaque; DE – 7,2; H – 2-2,5; CL – none; F – uneven; M – pseudo-cubic crystals, massive. It mainly occurs as paramorphs after argentite (high-temperature phase of the same composition).
Origin and occurrence: Hydrothermal in ore veins. Beautiful crystals over 50 mm (2 in) long occurred in the Himmelsfürst mine in Freiberg, in Annaberg and Schneeberg, Germany. Acicular crystals are known from Jáchymov, Czech Republic. It is common in association with silver, proustite, pyrargyrite, polybasite, stephanite, galena and other minerals in Mexico. Probably the best paramorphs after argentite up to 70 mm (2¾ in) across come from the Rayas mine, Guanajuato. Fine crystals occur in the Las Chispas mine, Arizpe, Sonora and many localities in Zacatecas, Chihuahua.
Application: important Ag ore.

Acanthite, 23 mm, Arizpe, Mexico

Argentite
Ag_2S

CUBIC ● ● ●

Properties: C – black-gray, tarnishes black; S – black; L – metallic; D – opaque; DE – 7,1; H – 2-2,5; CL – imperfect; F – uneven to conchoidal; M – octahedral and cubic crystals, dendritic aggregates, massive. Stabile at temperatures over 179EC (354.2°F), below this temperature there are paramorphs of acanthite after argentite.

Umangite, 40 mm, Beaverlodge Lake, Canada

Origin and occurrence: Hydrothermal in low-temperature ore deposits, associated with silver, galena and Ag sufosalts. Occurs between the oxidation and cementation zone with stromeyerite, silver, jalpaite, iodargyrite and other minerals. Fine crystals up to 40 mm (1 9/16 in) across, are known from Freiberg and Schneeberg, Germany. Similar crystals found in Jachymov and Midinec, Czech Republic. Crystals up to 30 mm (1 3/16 in) occur in Sarrabus, Sardinia, Italy. Maybe the best argentite crystals occurred in Mexico (Arizpe, Sonora; Zacatecas; Guanajuato), where crystals reached up to 40 mm (1 9/16 in). Fine crystals up to 20 mm (25/32 in) across reported from Chañarcillo in Chile.
Application: important Ag ore.

Argentite, 29 mm, Zacatecas, Mexico

Aguilarite, 4 mm xx, Guanajuato, Mexico

Hessite

Ag_2Te

MONOCLINIC ● ● ●

Properties: C – lead to steel-gray, tarnishes black; S – light gray; L – metallic; D – opaque; DE – 8,4; H – 2-3; CL – imperfect; F – even; M – pseudo-cubic crystal combinations, granular, massive.
Origin and occurrence: Hydrothermal in medium- and low-temperature ore veins together with calaverite, sylvanite, altaite, gold, tellurium and other sulfides. The best specimens with crystals up to 20 mm (25/32 in) across (25/32 in) and aggregates up to 100 mm (4 in) come from the Bote mine, Romania. Small crystals occur in the Jamestown mine, California, USA. Massive aggregates were found in Gold Hill, Colorado, USA and Moctezuma, Mexico. Aggregates, up to tens of cm^2 (several sq in) in size, were known in the Zavodinskii mine, Altai, Kazakhstan.

Aguilarite

Ag_4SeS

ORTHORHOMBIC ●

Properties: C – lead-gray, tarnishes black; S – gray-black; L – metallic; D – opaque; DE – 7,7; H – 2,5; CL – none; F – hackly; M – skeletal crystals, massive.
Origin and occurrence: Hydrothermal in ore veins, together with silver, stephanite, proustite, pearceite, calcite and quartz. The best crystallized specimens with crystals up to 30 mm (1 3/16 in) across come from the San Carlos mine, Guanajuato and Chontalpan, Taxco, Guerrero, Mexico. It is also known in inter-

Hessite, 31 mm, Botes, Romania

growths with acanthite and naumannite from the Comstock Lode, Virginia City, Nevada, USA.

Argyrodite

Ag_8GeS_6

ORTHORHOMBIC •

Properties: C – steel-gray, tarnishes black; S – gray-black; L – metallic; D – opaque; DE – 6,3; H – 2,5-3; CL – none; F – uneven to conchoidal; M – combinations of cubic crystals, botryoidal aggregates, massive.
Origin and occurrence: Hydrothermal in low-temperature base metal deposits, associated with Ag sulfosalts. It occurred in crystals in the Himmelsfhrst mine, Freiberg, Germany. Crystals were also found in several localities in Bolivia (Atoche, Colquechaca, Potosí). Crystal measuring 60 mm (24 in) across is reported from Porco, Bolivia.

Argyrodite, 34 mm, Mexico

Stromeyerite

AgCuS

ORTHORHOMBIC • • •

Properties: C – dark steel gray, tarnishes blue; S – steel-gray; L – metallic; D – opaque; DE – 6,3; H – 2,5-3; CL – none; F – conchoidal; M – pseudo-hexagonal tabular crystals, massive.
Origin and occurrence: Mostly secondary in the cementation zone of ore veins, associated with freibergite, bornite, chalcocite, galena and other minerals. Fine tabular crystals up to 10 mm ($^3/_8$ in) across found in Dzhezkazgan, Kazakhstan, where pseudo-morphs of stromeyerite after silver wires also occur. Skeletal prismatic pseudo-morphs after chalcocite crystals come from Vrancice, Czech Republic. Massive aggregates are common in many deposits in Colorado, USA (Aspen; Red Mountain), Chile (Copiapó), Bolivia (Potosí) and Canada (Cobalt, Ontario).

Stromeyerite, 6 mm xx, Dzhezkazgan, Kazakhstan

Jalpaite

Ag_3CuS_2

TETRAGONAL • •

Properties: C – light gray, tarnishes dark gray to iridescent; S – black; L – metallic; D – opaque; DE – 6,8; H – 2-2,5; CL – good; F – conchoidal; M – crystals, granular.
Origin and occurrence: Hydrothermal in low-temperature ore veins. Crystals up to 25 mm (1 in) across known from Jalpa, Querétaro, Mexico. Crystals up to 30 mm ($^{13}/_{16}$ in) across come from the Caribou mine, Colorado, USA. Massive aggregates, associated with galena, sphalerite, pyrite, stromeyerite, polybasite and other minerals, occurred in Príbram, Czech Republic.

Jalpaite, 30 mm veinlet, Príbram, Czech Republic

Pentlandite, 90 mm, Sudbury, Canada

Pentlandite

$(Fe,Ni)_9S_8$

CUBIC ● ● ● ●

Properties: C – light bronze to red-brown; S – light bronze; L – metallic; D – opaque; DE – 5,0; H – 3,5-4; CL – none; F – conchoidal; M – crystals, massive.
Origin and occurrence: Typical magmatic liquid mineral, associated with pyrrhotite and chalcopyrite. It is common in Sudbury, Ontario, Canada, as mostly microscopic inclusions in chalcopyrite, but also as imperfect crystals. Large accumulations are known from Talnakh near Norilsk, Siberia, Russia, where it occurs together with Cu, Pt and Pd sulfides. It is also important ore in the deposit near Rustenburg, South Africa.
Application: the most important Ni ore.

Sphalerite, 70 mm, Olkusz, Poland

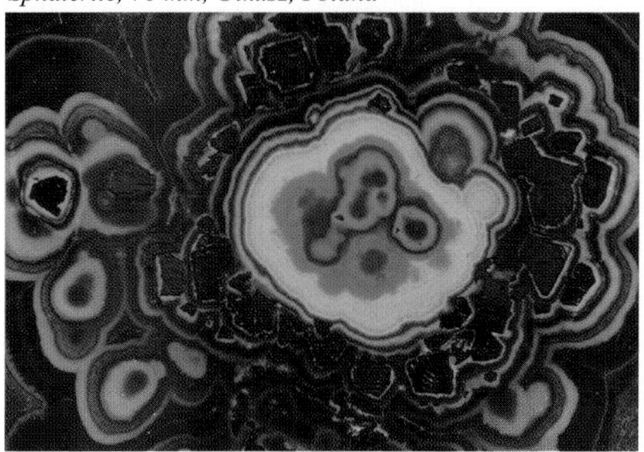

Sphalerite

ZnS

CUBIC ● ● ● ●

Varieties: cleiophane (green, yellow, orange), marmatite (black)

Properties: C – colorless, yellow, orange, green, brown, black; S – brownish, light yellow, white; L – resinous to adamantine; D – transparent, translucent, opaque; DE – 4,1; H – 3,5-4; CL – perfect; F – conchoidal; M – tetrahedral and dodecahedral crystals, botryoidal, fibrous and stalactitic aggregates, massive; LU – sometimes orange.
Origin and occurrence: Magmatic (liquid, in pegmatites); hydrothermal in low- to high-temperature deposits, skarns, hydrothermal sedimentary deposits;

Sphalerite, 31 mm, Elmwood, U.S.A.

rare sedimentary and metamorphic. It occurs together with galena, pyrite, chalcopyrite, marcasite, fluorite, barite, quartz and other minerals. Beautiful crystals up to 100 mm (4 in) across come from Trepéa, Serbia. Green and red crystals up to 100 mm (4 in) known from Cananea, Sonora, Mexico. Fine yellow crystals up to 30 mm (1$^{3}/_{16}$ in) were common in Banská Štiavnica, Slovakia; similar crystals occur in Madan, Bulgaria. The most beautiful yellow, orange and red crystals up to 150 mm (6 in) across found in Picos de Europa, Santander, Spain. They are sometimes faceted. Brown crystals up to 50 mm (2 in) come from Joplin, Missouri; stalactitic aggregates up to 150 mm (6 in) long come from Galena, Illinois, USA. Perfect black, shiny crystals and twins up to 50 mm (2 in) are famous from Dalnegorsk, Russia; yellow crystals, up to 30 mm (1$^{3}/_{16}$ in), occurred in Dzhezkazgan, Kazakhstan. Transparent crystals up to 30 mm (1$^{3}/_{16}$ in) also found in Franklin, New Jersey, USA. Green crystals up to 100 mm (approx. 4 in) across occurred in the Big Four mine, Colorado, USA. Crystals up to 50 mm (2 in) known from the Oppu mine, Aomori, Japan.
Application: principal Zn ore.

Coloradoite
HgTe

CUBIC ● ●

Properties: C – black-gray; S – black-gray; L – metallic; D – opaque; DE – 8,1; H – 2,5; CL – none; F – uneven to conchoidal; M – granular, massive.
Origin and occurrence: Hydrothermal, associated with altaite, calaverite, krennerite, gold, pyrite and other minerals in Au-bearing veins. It was common in Cripple Creek and in the Smuggler mine, Colorado; in the Norwegian mine, California, USA. Grains, reaching up to several mm, come from Jílové, Czech Republic.

Coloradoite, 70 mm, Kalgoorlie, Australia

Chalcopyrite, 10 mm xx, Cavnic, Romania

Chalcopyrite
$CuFeS_2$

TETRAGONAL ● ● ● ● ●

Properties: C – brass-yellow, tarnishes iridescent; S – green-black; L – metallic; D – opaque; DE – 4,3; H – 3,5-4; CL – imperfect; F – uneven; M – tetrahedral crystals, botryoidal aggregates, massive.
Origin and occurrence: Magmatic, hydrothermal and sedimentary, in association with sphalerite, galena, tetrahedrite, pyrite and many other sulfides. Fine crystals up to 30 mm (1$^{3}/_{16}$ in) across are known from Banská Stiavnica, Slovakia and from Cavnic, Romania. Crystals up to 120 mm (4$^{11}/_{16}$in) across, associated with other sulfides, come from the Nikolai mine in Dalnegorsk, Russia. Beautiful crystals up to 120 mm (4$^{11}/_{16}$in) found in Japan (Arawaka, Osarizawa). Fine crystals reaching up to several cm occur in Peru (Huanzala, Huaron). Massive aggregates are important Cu ore in Sudbury, Ontario, Canada; Bingham, Utah; Bisbee, Arizona, USA; and Rio Tinto, Spain.
Application: important Cu ore.

Chalcopyrite, 15 mm x, Ground Hog Mine, U.S.A.

Luzonite, 60 mm, Récsk, Hungary

Luzonite
Cu_3AsS_4

TETRAGONAL ● ●

Properties: C – dark pink-brown; S – black; L – metallic; D – opaque; DE – 4,5; H – 3,5; CL – good; F – uneven to conchoidal; M – crystals, granular, massive.
Origin and occurrence: Hydrothermal in low- to medium-temperature veins, associated with enargite, tetrahedrite, sphalerite, bismuthinite, Ag sulfosalts and other minerals. Crystals up to 40 mm (1 9/16 in) across come from Quiruvilca; it is common in Cerro de Pasco, Peru. It is also common in the Teine mine, Hokkaido, Japan; crystals also occur in Kinkwaseli, Taiwan. It is known from Bor, Serbia and Récsk, Hungary.

Stannite
Cu_2FeSnS_4

TETRAGONAL ● ● ●

Properties: : C – steel-gray to black, tarnishes blue;

Stannite, 10 mm xx, Potosí, Bolivia

Germanite, 40 mm, Tsumeb, Namibia

S – black; L – metallic; D – opaque; DE – 4,5; H – 4; CL – imperfect; F – uneven; M – pseudo-octahedral crystals, granular, massive.
Origin and occurrence: Hydrothermal in high-temperature Sn deposits. The best specimens come from Bolivia; crystals up to 50 mm (2 in) across known from Llallagua; crystals up to 30 mm (1 3/16 in) from Chocaya and cross-like inter-growths from the San José and Itos mines near Oruro. It occurs as massive vein fillings in Carn Brea, Cornwall, UK and Cínovec, Czech Republic. It was also found in amblygonite pegmatites in Caceres, Spain and in quartz-amblygonite veins near Vernérov, Czech Republic.
Application: Sn ore.

Germanite
$Cu_{26}Fe_4Ge_4S_{32}$

CUBIC ● ●

Properties: C– pink to purple-grey, S – black, L - metallic, D - opaque, DE – 4,5, H - 4, CL – none, F – uneven, M – tetrahedral crystals, granular, massive.
Origin and occurrence: Hydrothermal, often inter-grown with tennantite, bornite and other minerals. It was important only in one particular place in Tsumeb, Namibia, where larger accumulations were found. Smaller occurrences are known from the Shikanai mine, Japan; small cubic crystals come from the Humboldt mine, Colorado, USA.

Tennantite
$Cu_{12}As_4S_{13}$

CUBIC ● ● ● ●

Properties: C – steel-gray; S – black, brown to dark

red; L – metallic; D – opaque; DE – 4,6; H – 3-4,5; CL – none; F – conchoidal to uneven; M – tetrahedral crystals, granular, massive.
Origin and occurrence: Hydrothermal in ore veins and greisens with pyrite, calcite, dolomite, quartz and other sulfides and Cu-Pb-Zn-Ag sulfosalts. Fine crystals are known from Cornwall, UK (Wheal Jewel, Gwennap; Carn Brea). Crystals up to 150 mm (6 in) across, come from Tsumeb, Namibia. Crystals of binnite in Lengenbach, Binntal, Switzerland, up to 30 mm (1$^{3}/_{16}$ in). Crystals up to 20 mm ($^{25}/_{32}$ in) across were found recently in Dzhezkazgan, Kazakhstan. Large masses occurred in Kipushi, Zaire. Crystals up to 25 mm (1 in) known from El Cobre, Zacatecas, Mexico. *Application:* Cu ore.

Tennantite, 26 mm, El Cobre, Mexico

Tetrahedrite

$Cu_{12}Sb_4S_{13}$

CUBIC ● ● ● ●

Properties: C – steel-gray to black; S – black, brown; L – metallic; D – opaque; DE – 5,0; H – 3-4,5; CL – none; F – conchoidal; M – tetrahedral crystals, granular, massive.
Origin and occurrence: Hydrothermal in low- to medium temperature veins; in contact metamorphic deposits together with chalcopyrite, galena, sphalerite, pyrite, bornite, calcite, quartz and other minerals. The largest known crystals up to 25 cm (9$^{7}/_{8}$ in) across come from Anzen and Irazein in Pyrennees, France. Common crystals several cm in size occur in Cavnic, Romania. Fine specimens with crystals up to 70 mm (2¾ in) across found in the Mercedes mine in Huallanca, Peru. Other Peruvian localities like Casapalca and Morococha yielded fine crystallized specimens. Fine crystals up to 20 mm ($^{25}/_{32}$ in) in size known from Príbram, Czech Republic.
Application: Cu ore.

Tetrahedrite, 10 mm xx, Peru

Freibergite, 15 mm xx, Obecnice, Czech Republic

Freibergite

$(Ag,Cu,Fe)_{12}(Sb,As)_4S_{13}$

CUBIC ● ● ●

Properties: C – gray to black; S – black, brown to dark red; L – metallic; D – opaque; DE – 5,4; H – 3-4,5; CL – none; F – uneven to conchoidal; M – tetrahedral crystals, massive.
Origin and occurrence: Hydrothermal in ore deposits, associated with many sulfides and sulfosalts. It is mainly massive in the Himmelsfürst mine in Freiberg, Germany; Cobalt, Ontario, Canada and in Mount Isa, Queensland, Australia. It was a principal Ag ore in Kutná Hora, Czech Republic.
Application: Ag ore.

Wurtzite

ZnS

HEXAGONAL ● ● ●

Properties: C – dark red-brown, dark brown to brown-black; S – brown; L – resinous to submetallic; D – translucent to opaque; DE – 4,1; H – 3,5-4; CL – good; M – pyramidal, prismatic to thick tabular, striated crystals, concentric banded and radial aggregates.
Origin and occurrence: Hydrothermal in ore veins with sphalerite, marcasite, pyrite and other minerals. Also of low-temperature origin along the cracks of clay concretions. The best wurtzite crystals come from Bolivia; crystals up to 40 mm (1 9/16 in) across from Animas and crystals up to 20 mm (25/32 in) across from Llallagua and Potosí. Fine crystals up to 30 mm (1 3/16 in) across found in Talnakh near Norilsk, Siberia, Russia. Interesting radial aggregates, up to several cm in diameter, occurred in Príbram, Czech Republic.

Wurtzite, 30 mm xx, Animas, Bolivia

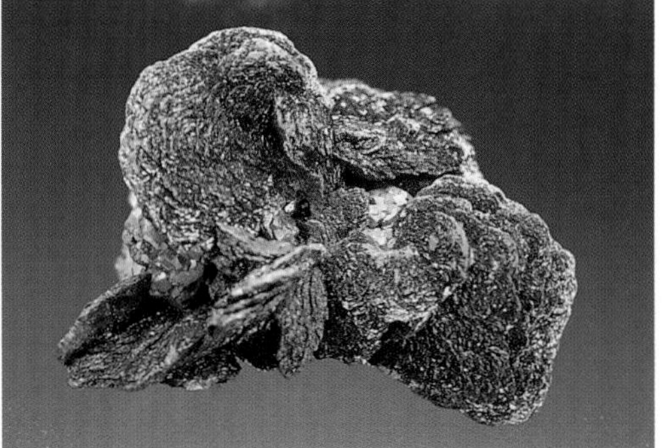

Greenockite

CdS

HEXAGONAL ● ● ●

Properties: C – yellow to orange, dark red; S – orange-yellow to brick-red; L – adamantine to resinous; D – opaque to translucent; DE – 4,8; H – 3-3,5; CL – good; F – conchoidal; M – trillings, pulverulent coatings.
Origin and occurrence: Mainly secondary as pulverulent coatings on sphalerites. It also occurs in cavities of volcanic rocks together with prehnite, zeolites and calcite. Crystals are known from ore veins in Llallagua, Bolivia. Pulverulent coatings are described from Príbram, Czech Republic; Bleiberg, Austria; and from the deposits in the Tri State region, Missouri, USA. Crystals up to 10 mm (3/8 in) occurred in the cavities of volcanic rocks near Renfrew, Scotland, UK.

Enargite

Cu_3AsS_4

ORTHORHOMBIC ● ● ● ●

Properties: C – gray-black to black; S – gray-black;

Grenockite, 20 mm, Ocna de Fier, Romania

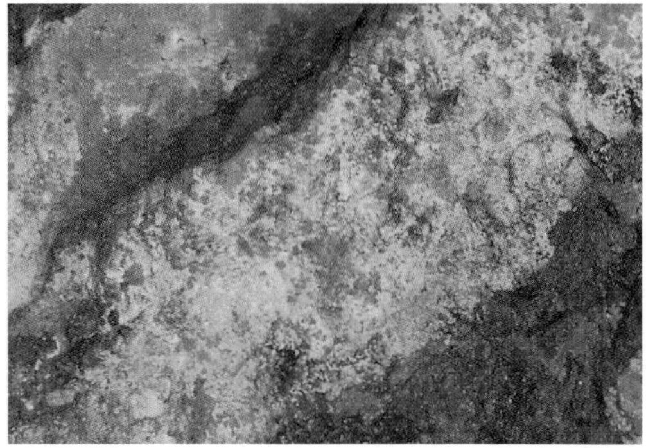

L – metallic; D – opaque; DE – 4,4; H – 3; CL – perfect; F – uneven; M – tabular and prismatic, striated crystals, massive.
Origin and occurrence: Hydrothermal in medium-temperature, sometimes in low-temperature deposits, associated with quartz, pyrite, sphalerite, galena, bornite and other minerals. Beautiful crystals, up to 100 mm (4 in) across, come from the Luz Angelica mine in Quiruvilca; crystals up to 150 mm (6 in) across from Morococha and Cerro de Pasco, Peru. Fine crystals were found in Butte, Montana, USA and in Mancayano, Luzon, Philippines. It occurs as a principal Cu ore in several deposits (Bor, Serbia; Huaron, Peru).
Application: Cu ore.

Enargite, 41 mm, Butte, U.S.A.

Cubanite

Cu_2FeS_3

ORTHORHOMBIC ● ● ●

Properties: C – brass-yellow to bronze; S – black; L – metallic; D – opaque; DE – 4,1; H – 3,5; CL – none; F – conchoidal; M – tabular crystals, massive.
Origin and occurrence: Magmatic in liquid deposits as inclusions in chalcopyrite, as hydrothermal in high-temperature deposits, associated with chalcopyrite, pyrite, pyrrhotite and sphalerite. The best crystals, twins, up to 40 mm ($1\frac{9}{16}$ in) across, come from the Henderson No.2 mine, Chibougamau, Quebec, Canada. It was also described in crystals from Sudbury, Ontario, Canada and Morro Velho, Brazil. Its large accumulations are important Cu ore in Sudbury, Ontario, Canada; in Mooihoek, South Africa and in Prince William Sound, Alaska, USA.
Application: important Cu ore.

Cubanite, 33 mm x, Chibougamau, Canada

Sternbergite, 20 mm, St.Andreasberg, Germany

Sternbergite

Ag_2FeS_3

ORTHORHOMBIC ● ●

Properties: C – golden-brown; S – black; L – metallic to adamantine; D – opaque; DE – 4,3; H – 1-1,5; CL – perfect; M – thin tabular pseudo-hexagonal crystals, often in rosettes and fan-shaped aggregates. *Origin and occurrence:* Hydrothermal in Ag-bearing veins, associated with stephanite, acanthite, proustite, argentopyrite and other minerals. Tabular crystals up to several mm across known from Jáchymov and Medenec, Czech Republic; from St. Andreasberg, Johanngeorgenstadt, Schneeberg and Freiberg, Germany.

Pyrrhotite, 68 mm, Dalnegorsk, Russia

Argentopyrite, 60 mm, Medenec, Czech Republic

Argentopyrite

Ag_2FeS_3

ORTHORHOMBIC ● ●

Properties: C – gray-white, tarnishes iridescent; S – gray; L – metallic; D – opaque; DE – 4,3; H – 3,5 – 4; CL – none; F – uneven; M – thick tabular pseudo-hexagonal crystals, granular.
Origin and occurrence: Hydrothermal in ore veins, associated with proustite, pyrargyrite, stephanite, sternbergite, dolomite, quartz and other minerals. Crystals up to 5 mm ($^{3}/_{16}$ in) across come from Jáchymov and Medenec, Czech Republic; it is also known from Schlema, Freiberg; Schneeberg, Germany; and from Colquechaca, Bolivia.

Pyrrhotite

$Fe_{1-x}S$ (x = 0-0,17)

MONOCLINIC ● ● ● ●

Properties: C – bronze-yellow to brown, tarnishes quickly; S – dark gray-black; L – metallic; D – opaque; DE – 4,7; H – 3,5-4,5; CL – none; F – uneven to conchoidal; M – tabular, pyramidal and prismatic crystals, massive.
Origin and occurrence: Magmatic liquid in basic rocks, together with pyrite and pentlandite; in pegmatites; hydrothermal in high-temperature and metasomatic deposits; sedimentary and metamorphic. Tabular crystals up to 300 mm ($11^{13}/_{16}$ in) across come from Trepéa, Serbia and Dalnegorsk, Russia; prismatic crystals up to 150 mm (6 in) long found in Santa Eulalia, Chihuahua, Mexico and Chiuzbaia, Romania. Tabular crystals up to 110 mm ($4^{5}/_{16}$ in) occurred in Cavnic, Romania. Large imperfect crystals, coated with wavellite, are known from Llallagua, Bolivia. Huge masses of industrial importance occur in Sudbury, Ontario, Canada; Talnakh near Norilsk, Siberia, Russia and elsewhere. *Application:* sometimes as Fe ore.

Nickeline, 138 mm, Pöhla, Germany

Nickeline
NiAs

HEXAGONAL ● ● ● ●

Properties: C – light copper-red, tarnishes gray to black; S – light brown-black; L – metallic; D – opaque; DE – 7,8; H – 5-5,5; CL – none; F – conchoidal; M – striated crystals, botryoidal and dendritic aggregates, granular, massive.
Origin and occurrence: Hydrothermal in high-temperature ore veins with skutterudite, nickel-skutterudite, safflorite, rammelsbergite and other minerals. Also magmatic in norites and peridotites. Crystals up to 15 mm ($^{19}/_{32}$ in) across come from Pöhla; small crystals were found in Richelsdorf, Germany. Fine massive aggregates and small crystals occurred in Bou Azzer, Morocco. Huge accumulations are known in Cobalt and Gowganda, Ontario, Canada.
Application: Ni ore.

Breithauptite
NiSb

HEXAGONAL ● ●

Properties: C – light copper-red, purplish; S – red-brown; L – metallic; D – opaque; DE – 8,6; H – 5,5; CL – none; F – conchoidal to uneven; M – thin tabular crystals, dendritic aggregates, massive.
Origin and occurrence: Magmatic, hydrothermal and metamorphic, associated with silver, nickeline, cobaltite and other sulfides. It occurs inter-grown with pyrrhotite and pentlandite in magmatic liquid deposit Vlakfontein, South Africa. It is common in hydrothermal veins in Cobalt, Ontario, Canada; also known from Sarrabus, Sardinia, Italy; and St. Andreasberg, Germany.

Breithauptite, 80 mm, Cobalt, Canada

Millerite, 26 mm, Antwerp, U.S.A.

Millerite

NiS

TRIGONAL ● ● ●

Properties: C – light brass-yellow to bronze, tarnishes iridescent; S – greenish-black; L – metallic; D – opaque; DE – 5,4; H – 3-3,5; CL – perfect; F – uneven; M – acicular crystals, cleavable masses, aggregates of parallel inter-grown crystals with velvety surface.
Origin and occurrence: Hydrothermal in low-temperature ore deposits, also as a product of decomposition of Ni sulfides. It is associated with pyrrhotite, ankerite, whewellite, barite and other minerals. Cleavable masses occurred near Temagami near Sudbury, Ontario, Canada and in Kambalda, Western Australia. Acicular crystals come from ore veins in Jáchymov and Príbram, Czech Republic; needles in cavities of siderite concretions in the vicinity of Kladno, Czech Republic, reached up to 70 mm (2¾ in). Acicular crystals up to 80 mm (3⅛ in) long known from the limestone cavities near Dortmund, Germany. Beautiful specimens with crystals up to 50 mm (2 in) long in the cavities in hematite occurred in the Sterling mine, Antwerp, New York, USA. Velvety fibrous aggregates were found in the Thompson mine, Manitoba, Canada. *Application:* Ni ore.

Alabandite, 40 mm, Sacarimb, Romania

Alabandite

MnS

CUBIC ● ●

Properties: C – black; S – green; L – submetallic; D – opaque; DE – 4,1; H – 3,5-4; CL – perfect; F – uneven; M – cubic and octahedral crystals, granular, massive.
Origin and occurrence: Hydrothermal in ore veins, associated with rhodochrosite, calcite, galena, sphalerite, pyrite and other minerals. Crystals up to 20 mm (25/32 in) across together with granular aggregates are relatively common in Romania (Săcărîmb, Baia de Arieş, Roşia Montana). Crystals are also known from the Queen of the West mine, Colorado and from the Lucky Cuss mine, Tombstone, Arizona, USA.

Galena, 56 mm, Neudorf, Germany

Galena, 66 mm, Picher, U.S.A.

Galena

PbS

CUBIC ● ● ● ● ●

Varieties: steinmannite

Properties: C – lead-gray; S – lead-gray; L – metallic; D – opaque; DE – 7,6; H – 2,5; CL – perfect; F – conchoidal; M – cubic crystals and their complex combinations, tabular crystals, skeletal aggregates, massive.

Origin and occurrence: Magmatic, hydrothermal, metamorphic, very rare sedimentary, associated with sphalerite, chalcopyrite, pyrite, quartz and other minerals. Large crystals up to several tens of cm in size come from many localities in the USA. (Joplin, Missouri; Galena, Kansas; Picher, Oklahoma; Sweetwater mine, Missouri). Beautiful, often skeletal crystals or spinel-law twins up to 200 mm ($7^7/_8$ in) across known from the Nikolai mine in Dalnegorsk, Russia. Fine crystals occur also in Naica, Chihuahua, Mexico. Famous complicated combinations of crystals were found in Neudorf, Germany. Octahedral crystals up to 10 mm ($^3/_8$ in) (steinmannite variety) were common in Příbram, Czech Republic. Beautiful specimens with cubes up to several cm across come from Madan, Bulgaria; spinel-law twins occurred in Herja, Romania.

Application: the most important Pb ore.

Clausthalite, 20 mm veinlet, Tilkerode, Germany

Clausthalite

PbSe

CUBIC ● ● ●

Properties: C – lead-gray, bluish; S – gray-black; L – metallic; D – opaque; DE – 8,3; H – 2,5 – 3; CL – good; F – granular; M – granular, massive..

Origin and occurrence: Hydrothermal in ore veins with a low S content, together with berzelianite, umangite, uraninite and other minerals. It occurs as massive aggregates in calcite veins with other selenides in Clausthal and Tilkerode, Germany. It is similar in Skrikerum, Sweden; common in the U deposits in Predborice, Bukov and Zlatkov, Czech Republic.

Altaite, 65 mm, Stania, Romania

Altaite

PbTe

CUBIC ● ● ●

Properties: C – tin-white to yellowish, tarnishing to bronze; S – black; L – metallic; D – opaque; DE – 8,3; H – 3; CL – perfect; F – conchoidal; M – cubic and octahedral crystals, massive.

Origin and occurrence: Hydrothermal in vein Au deposits, associated with other tellurides, galena and other minerals. Crystals up to 20 mm (25/32 in) across are known from the Revenge mine, Colorado, USA. It is massive with aguilarite in Kalgoorlie, Western Australia, Australia. It is relatively common with other tellurides in Sâcârîmb, Romania; in the Zavodinskii mine, Altai, Russia; and in Zod near Sevan Lake, Armenia.

Miargyrite

$AgSbS_2$

MONOCLINIC ● ● ●

Properties: : C – black to steel-gray; S – cherry-red; L – metallic, adamantine; D – opaque; DE – 5,3; H – 2,5; CL – imperfect; F – conchoidal; M – thick tabular, striated crystals, massive.
Origin and occurrence: Hydrothermal in low-temperature ore veins, together with proustite, pyrargyrite, polybasite, silver, quartz and other minerals. Fine crystals, up to 10 mm (3/8 in) across, occurred in Príbram and Kutná Hora, Czech Republic; in Bräunsdorf and Freiberg, Germany. It is also known from many localities in Bolivia (Tatasi, Oruro – 10 mm (3/8 in) crystals, Potosí) and Mexico (Sombrerete, Catorce). It was also found in Hiendelaencina, Guadalajara, Spain.
Application: Ag ore.

Franckeite

$(Pb,Sn)_6FeSn_2Sb_2S_{14}$

TRICLINIC ● ●

Properties: C – gray-black; S – gray-black; L – metallic; D – opaque; DE – 5,9; H – 2,5-3; CL –

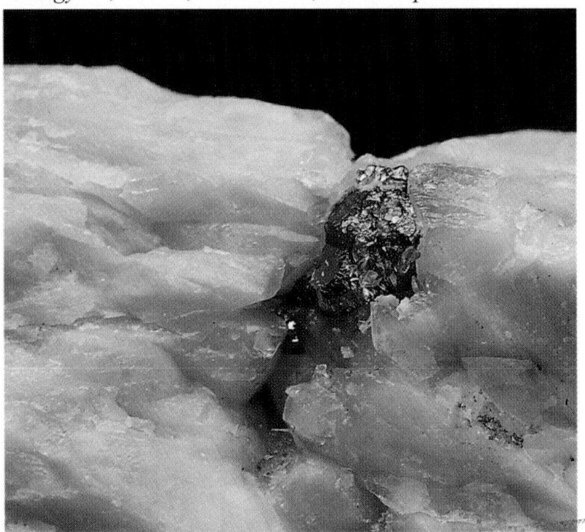

Miargyrite, 20 mm, Kutná Hora, Czech Republic

Franckéite, 30 mm, Oruro, Bolivia

Cylindrite, 50 mm, Poopó, Bolivia

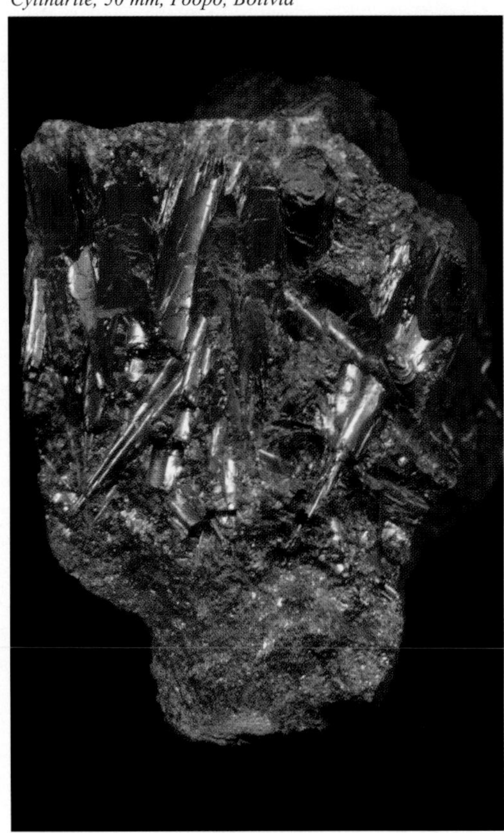

perfect; M – thin tabular, curved crystals, often rosette-like aggregates of crystals.
Origin and occurrence: Hydrothermal in Ag-Sn deposits, associated with cylindrite, zinkenite, cassiterite, wurtzite and other minerals; in contact metamorphic limestones. The best crystals up to 60 mm (24 in) across come from various localities in Bolivia (Mollecagua near Poopó; Huanuni; Chocoya near Potosí; and Colquechaca and Oruro). It is known in contact metamorphic limestone from the Kalkar quarry, California, USA.
Application: Sn ore.

Cylindrite
$Pb_3Sn_4FeSb_2S_{14}$

TRICLINIC ● ●

Properties: C – lead-gray to black; S – black; L – metallic; D – opaque; DE – 5,4; H – 2,5; CL – perfect; M – cylindrical, conical and spherical crystal aggregates, massive.
Origin and occurrence: Hydrothermal in Sn-bearing veins with franckeite, stannite, cassiterite, galena and other minerals. World famous specimens with crystals up to 50 mm (2 in) long come from the Trinacria and Santa Cruz mines near Poopó, Bolivia. It was also reported from the Smirnovsk deposit, Transbaikalia, Russia.

Cinnabar
HgS

TRIGONAL ● ● ● ●

Properties: C – red to brownish-red; S – crimson; L – adamantine to metallic, also dull; D – opaque; DE – 8,2; H – 2-2,5; CL – perfect; F – conchoidal to uneven; M – rhombohedral, thick tabular and prismatic crystals, massive.
Origin and occurrence: Low-temperature hydrothermal mineral, associated with realgar, mercury, pyrite, marcasite and other minerals. The world's best specimens with crystals up to 70 mm (2¾ in) across are known from many localities in China (Hunan and Guizhou provinces). Fine shiny crystals up to 10 mm (3/8 in) across come from Nikitovka, Ukraine and in Khaidarkan, Kyrgyzstan. Crystals up to 30 mm (1 3/16 in) also found in Monte Amiata and Rippa near Seravezza, Italy. Its massive aggregates are an important Hg ore in Allchar, Macedonia; Almaden, Spain and elsewhere.
Application: the most important Hg ore.

Cinnabar, 30 mm x, Hunan, China

Covellite, 65 mm, Butte, U.S.A.

Covellite
CuS

HEXAGONAL ● ● ●

Properties: C – indigo-blue, tarnishing iridescent; S – lead-gray; L – submetallic to resinous; D – opaque; DE – 4,6; H – 1,5-2; CL – perfect; F – uneven; M – hexagonal tabular crystals, massive.
Origin and occurrence: Rare hydrothermal; mainly secondary in the oxidation zone of ore deposits, associated with chalcopyrite, chalcocite, djurleite, bornite and other minerals. Tabular crystals up to 30 mm (1 3/16 in) across are known from Butte, Montana and Summitville, Colorado, USA; also from Sarrabus, Sardinia, Italy. Massive aggregates are common in Bor, Serbia; Bisbee, Arizona, USA and elsewhere.
Application: Cu ore.

Linneite
$Co^{2+}Co^{3+}{}_2S_4$

CUBIC ● ● ●

Properties: C – light gray to steel-gray; S – black-gray; L – metallic; D – opaque; DE – 4,9; H – 4,5-5,5; CL – imperfect; F – uneven to conchoidal; M – octahedral crystals, granular, massive.
Origin and occurrence: Hydrothermal in ore veins and metamorphic deposits, together with chalcopyrite, pyrrhotite, millerite, bismuthinite and sphalerite. The best crystals up to 30 mm (1 3/16 in) across come from the Kilembe mine, Uganda. Crystals are also known from Musonoi, Shaba, Zaire; Mhsen, Germany; siderite concretions in Kladno, Czech Republic and from the Bastnäs mine near Riddarhyttan, Sweden.

Carrollite
$Cu(Co,Ni)_2S_4$

CUBIC ● ● ●

Properties: C – light to steel-gray, tarnishing red-purple; S – gray; L – metallic; D – opaque; DE – 4,8; H – 4,5-5,5; CL – imperfect; F – conchoidal to uneven; M – octahedral crystals, granular, massive.
Origin and occurrence: Hydrothermal in ore veins, associated with linnéite, chalcopyrite and other minerals. It is a principal Co ore in deposits in Zaire. Beautiful crystals up to 20 mm (25/32 in) across are known from the M'Sesa mine near Kambove, Kolwezi and from Kamoto, Shaba.
Application: Co ore.

Linnéite, 60 mm, Ruwenzori, Uganda

Carrollite, 8 mm x, Kambove, Zair

Stibnite

Sb_2S_3

ORTHORHOMBIC ● ● ● ●

Properties: C – steel-gray, tarnishing iridescent or black; S – lead-gray; L – metallic; D – opaque; DE – 4,6; H – 2; CL – perfect; F – conchoidal to uneven; M – thick to thin prismatic crystals, thin needles, massive.

Origin and occurrence: Hydrothermal in medium- and low-temperature ore veins with quartz and gold, the other associated minerals are rare (arsenopyrite, berthierite, gudmundite, antimony). The largest known stibnite crystals occurred in the in the Ichinokawa mine, Shikoku, Japan, where they were up to 60 cm (24 in) long. Similar sized crystals were found recently in several localities in China (the Xikuangshan mine, Hunan). Crystals up to 200 mm ($7^{7}/_{8}$ in) across come from Manhattan, Nevada, USA. Beautiful crystals are known from Romania; long prismatic ones with barite crystals are prevalent in Baia Sprie, clusters of thin needles come from Herja and thick prismatic crystals from Baiuţ. Perfect druses of crystals up to 150 mm (6 in) long, associated with purple fluorite crystals, barite and calcite, occurred in Kadamdzhai, Kyrgyzstan. Beautiful druses of stibnite crystals with quartz come also from Kremnica, Slovakia. Also Kostajnik, Serbia and La Lucette, France yielded fine crystals in the past.

Application: Sb ore.

Stibnite, 187 mm, Ichinokawa, Japan

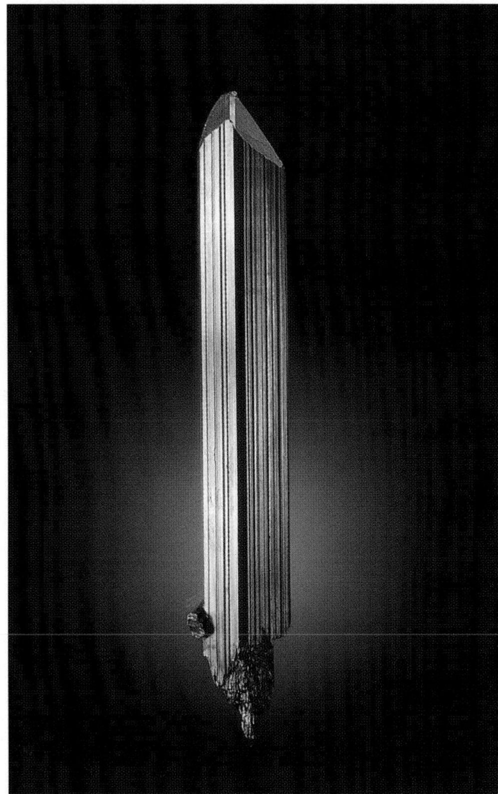

Stibnite, 120 mm, Baia Sprie, Romania

Bismuthinite, 160 mm, Tasna, Bolivia

Tetradymite, 80 mm, Ciclova, Romania

Bismuthinite

Bi_2S_3

ORTHORHOMBIC ● ● ● ●

Properties: C – lead-gray to tin-white, tarnishing yellow and iridescent; S – lead-gray; L – metallic; D – opaque; DE – 6,8; H – 2-2,5; CL – perfect; F – uneven; M – thick prismatic to acicular, striated crystals, fibrous aggregates, massive.
Origin and occurrence: Hydrothermal in ore deposits; also in recent volcanic exhalation deposits, associated with bismuth, arsenopyrite, stannite, galena and other minerals. The world's best crystals over 50 mm (2 in) long, come from Bolivia (Tasna; Huanuni; Llallagua). Fine crystals were also found in Redruth, Cornwall, UK. Rich finds were made in Biggenden, Queensland, Australia. Interesting crystals are known from Spind, Norway.
Application: Bi ore.

Kermesite, 20 mm xx, Freiberg, Germany

Kermesite

Sb_2S_2O

TRICLINIC ● ● ●

Properties: C – cherry-red; S – brown-red; L – adamantine to submetallic; D – translucent to opaque; DE – 4,7; H – 1-1,5; CL – perfect; M – acicular crystals, radial aggregates.
Origin and occurrence: Secondary, as a result of a stibnite oxidation in Sb deposits, associated with stibnite, antimony, senarmontite, valentinite and stibiconite. Famous specimens with needles up to 100 mm (4 in) long, in radial aggregates, come from Pezinok and Pernek, Slovakia. Crystals up to 50 mm (2 in) long found in the Globe and Phoenix mines, Zimbabwe. It is also known from Bolivia (San Francisco mine, Poopó; Oruro) and Bräunsdorf, Germany.

Nagyagite, 100 mm, Sacarimb, Romania

Tetradymite

Bi_2Te_2S

TRIGONAL ● ● ●

Properties: C – light steel-gray, tarnishing dark to iridescent; S – light steel-gray; L – metallic; D – opaque; DE – 7,3; H – 1,5-2; CL – perfect; M – pyramidal crystals, curved bladed aggregates, - massive.
Origin and occurrence: Hydrothermal in medium- and high-temperature Au deposits; also in contact metamorphosed deposits, together with gold, hessite, calaverite, pyrite and other minerals. Crystals up to 10 mm (3/8 in) across are known from Župkov, Slovakia. It is common in Colorado (Red Cloud) and in California (Carson Hill), USA. It occurs with gold in skarns in Baiţa Bihorului, Romania.

Nagyagite

$Pb_5Au(Te,Sb)_4S_{5\text{-}8}$

TETRAGONAL ● ●

Properties: C – black-gray; S – black-gray; L – metallic; D – opaque; DE – 7,5; H – 1,5; CL – perfect; F – uneven; M – thin tabular to foliated crystals, granular, massive.
Origin and occurrence: Hydrothermal in low-temperature veins, associated with altaite, arsenic, gold, rodochrosite and other minerals. Foliated crystals up to 40 mm (1 9/16 in) across come from Săcărîmb and Baia de Arieş , Romania. Fine specimens were also found in Tavua, Viti Levu, Fiji and in the Sylvia mine, Tararu Creek, New Zealand. It also occurred in Gold Hill and Cripple Creek, Colorado, USA and in Kalgoorlie, Western Australia, Australia.

Sylvanite

$(Au,Ag)_2Te_4$

MONOCLINIC ● ● ●

Properties: C – steel-gray to silver-white, tarnishing to yellow; S – steel-gray to silver- white; L – metallic; D – opaque; DE – 8,2; H – 1,5-2; CL – perfect; F – uneven; M – short prismatic to thick tabular crystals, skeletal aggregates, granular.
Origin and occurrence: Hydrothermal in low-temperature ore veins, also in medium- and high-temperature deposits as one of the latest minerals, associated with gold, calaverite, hessite, krennerite and other minerals. The best specimens come from Săcărîmb and Baia de Arieş, Romania, where it occurs as skeletal crystals and aggregates. Crystals were also found in Cripple Creek, Colorado, USA; crystals up to 10 mm (3/8 in) across come from the Emperor mine, Viti Levu, Fiji.

Sylvanite, 20 mm, Cripple Creek, U.S.A.

Krennerite, 3 mm xx, Sacarimb, Romania

Krennerite

$AuTe_2$

ORTHORHOMBIC ● ●

Properties: C – silver-white to brass-yellow; S – silver-white; L – metallic; D – opaque; DE – 8,9; H – 2-3; CL – perfect; F – conchoidal to uneven; M – short prismatic, striated crystals, massive.
Origin and occurrence: Hydrothermal in gold-bearing veins, associated with gold, tellurium, pyrite, quartz and other tellurides. Crystals up to 20 mm ($^{25}/_{32}$ in) across found in Tavua, Viti Levu, Fiji. Small crystals occurred in Săcărîmb and Baia de Arieş, Romania and in Cripple Creek, Colorado, USA.
Application: Au ore.

Calaverite

$AuTe_2$

MONOCLINIC ● ● ●

Properties: C – brass-yellow to silver-white; S – greenish; L – metallic; D – opaque; DE – 9,3; H – 2,5-3; CL – none; F – uneven to conchoidal; M – bladed and short prismatic, striated crystals, granular, massive.
Origin and occurrence: Hydrothermal in low-temperature Au-bearing veins, sometimes in medium- and high-temperature deposits, associated with coloradoite, altaite, krennerite and other tellurides. It is common in the Mother Lode in California (Carson Hill). It is important, together with hessite in Cripple Creek, crystals up to 10 mm ($^{3}/_{8}$ in) across come from the Cresson mine, Colorado, USA. It is also known from several mines near Kirkland Lake, Ontario, Canada.
Application: Au ore.

Pyrite

FeS_2

CUBIC ● ● ● ● ●

Properties: C – light brass-yellow, tarnishing iridescent and darkens; S – green-black to brown-black; L – metallic; D – opaque; DE – 5,0; H – 6-6,5; CL – imperfect; F – conchoidal to uneven; M – combinations of cubic crystals, striated, stalactitic and spherical aggregates, massive.
Origin and occurrence: Magmatic segregations in basic rocks with pyrrhotite and pentlandite, in pegmatites and skarns; hydrothermal in porphyry and vein deposits together with other sulfides; hydrothermal sedimentary, sedimentary and metamorphic. Magmatic segregations are known from Sudbury, Ontario, Canada and Merensky Reef, Transvaal, South Africa. Large crystals up to 200 mm ($7^{7}/_{8}$ in) across are known from Rio Marina, Elba, Italy. Fine octahedra come from Llallagua, Bolivia. Crystals, up to 120 mm ($4^{11}/_{16}$in) across, were found in Bingham and Park City, Utah, USA. Crystals up to 150 mm (6 in) are known from Huanzala and Quiruvilca, Peru. The largest pyrite deposit is Rio Tinto, Spain, where fine-grained pyrite formed accumulations about 1 billion tons. Beautiful cubes up to 80 mm ($3^{1}/_{8}$ in) come from Navajún, Spain. Large deformed crystals

Calaverite, 21 mm, Nagyag, Romania

Pyrite, 74 mm, Navajún, Spain

Pyrite, 135 mm, Washington, U.S.A.

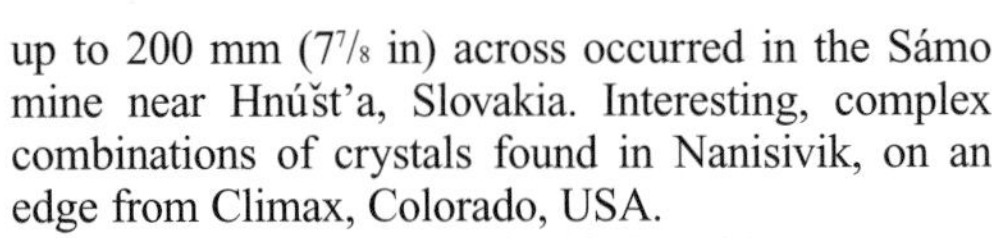

up to 200 mm ($7\frac{7}{8}$ in) across occurred in the Sámo mine near Hnúšt'a, Slovakia. Interesting, complex combinations of crystals found in Nanisivik, on an edge from Climax, Colorado, USA.
Application: production of sulfuric acid.

Hauerite

MnS_2

CUBIC ● ● ●

Properties: C – red-brown to brown-black; S – brown-red; L – metallic to adamantine; D – opaque; DE – 3,44; H – 4; CL – perfect; F – uneven to conchoidal; M – octahedral crystals and their combinations, spherical aggregates.
Origin and occurrence: Low-temperature sedimentary mineral, limited to clays with high S contents. The best crystals up to 50 mm (2 in) across come from the Destricella mine near Raddusa, Sicily, Italy. Crystals up to 25 mm (1 in) and their aggregates are known from Víglašská Huta (former Kalinka) near Zvolen, Slovakia. Crystals up to 15 mm ($^{19}/_{32}$ in) across occur in Tarnobrzeg, Poland. Also found with sulfur, gypsum, realgar and calcite in the salt domes near High Island, Texas, USA.

Hauerite, 39 mm, Raddusa, Italy

Sperrylite, 55 mm, Talnakh, Russia

Sperrylite

$PtAs_2$

CUBIC ● ●

Properties: C – tin-white; S – black; L – metallic; D – opaque; DE – 10,8; H – 6-7; CL – imperfect; F – conchoidal; M – complex combinations of cubic crystals, massive.
Origin and occurrence: It is mainly of magmatic liquid origin, associated with pyrrhotite, pentlandite, cubanite and other minerals. The best crystals come from Talnakh near Norilsk, Siberia, Russia, where inter-grown crystals reached up to 50 mm (2 in). Crystals up to 40 mm (1$^{9}/_{16}$ in) across known from Tweefontein, Potgietersrust, Bushveld, South Africa. It also occurred as disseminated aggregates in Sudbury, Ontario, Canada.

Aurostibite

$AuSb_2$

CUBIC ● ●

Properties: C – gray, tarnishing iridescent; S – gray; L – metallic; D – opaque; DE – 9,9; H – 3; CL Bnone; M – granular.
Origin and occurrence: Hydrothermal in quartz veins, associated with gold and other Sb minerals. Grains up to 5 mm ($^{3}/_{16}$ in) across come from Krásná Hora, Czech Republic. It also occurs in the Giant Yellowknife mine, Northwest Territories and Hemlo, Thunder Bay, Ontario, Canada and is also reported from Bestyube, Kazakhstan.

Krutaite

$CuSe_2$

CUBIC ●

Properties: C – gray; S – gray; L – metallic; D – opaque; DE – 6,5; H – 4; M – microscopic crystals, massive.
Origin and occurrence: Hydrothermal, associated with clausthalite, uraninite and other minerals. The richest accumulations of massive aggregates were found in the El Dragón mine, Potosí, Bolivia, where crystals up to 1 mm ($^{1}/_{32}$ in) occurred as well. It was originally described as microscopic from Petrovice, Czech Republic.

Cobaltite

CoAsS

ORTHORHOMBIC ● ● ●

Properties: C – silver-white; S – gray-black; L – metallic, adamantine to dull; D – opaque;DE – 6,3; H – 5,5; CL – perfect; F – uneven; M – pseudo-cubic crystals, granular, massive.
Origin and occurrence: Hydrothermal in high-temperature ore deposits and metamorphic, together

Aurostibite, 4 mm grain, Krásná Hora, Czech Republic

Krutaite, 40 mm, El Dragon, Bolivia

Cobaltite, 5 mm x, Hakansboda, Sweden

with magnetite, sphalerite, chalcopyrite, other sulfides and arsenides of Co and Ni. The best dodecahedra, several cm across, come from metamorphic sulfide deposits in Tunaberg, Sweden. Other Swedish localities, like Håkansboda and Ramsberg, yielded crystals up to 60 mm (24 in) across. Fine crystals about 10 mm (3/8 in) occurred in the magnesite deposit Mútnik near Hnúšt'a, Slovakia. Cubic crystals up to 30 mm (1 3/16 in) across found in Espanola, Ontario, Canada. Beautiful crystals are known from the skarn in Bimbowrie, South Australia, Australia.
Application: Co ore.

Gersdorfite

NiAsS

CUBIC ● ● ●

Properties: C – silver-white to steel-gray, tarnishing gray-black; S – gray-black; L – metallic; D – opaque; DE – 6,0; H – 5,5; CL – perfect; F – uneven; M – octahedral striated crystals and their combinations, massive. *Origin and occurrence:* Hydrothermal in medium-temperature ore deposits in association with nickeline, nickel-skutterudite, ullmanite, siderite and other minerals. Crystals up to 100 mm (4 in) across come from the Snowbird mine, Montana, USA. It is important in several mines near Sudbury and Cobalt, Ontario, Canada. Large cleavable masses are known from Dobšiná, Rudnany and Nizná Slaná, Slovakia. Crystals from the Aït Ahmane mine near Bou Azzer, Morocco, reached up to 40 mm (1 9/16 in) . *Application:* Ni ore.

Gersdorffite, 10 mm xx, Bou Azzer, Morocco

Ullmanite, 60 mm, Bou Cricha, Morocco

Ullmanite

NiSbS

CUBIC ● ●

Properties: C – steel-gray to silver-white; S – gray-black; L – metallic; D – opaque; DE – 6,8; H – 5-5,5; CL – perfect; F – uneven; M – combinations of cubic crystals, massive.

Origin and occurrence: Hydrothermal in ore veins together with other Ni minerals. Fine crystals, up to 20 mm (25/32 in) across, come from Monte Narba near Sarrabus, Sardinia, Italy. Twins are known from Lölling, Austria. Crystals up to 10 mm (3/8 in) were found in Kšice near Strîbro, Czech Republic. It is common in Broken Hill, New South Wales, Australia and in Cochabamba, Bolivia.

Application: Ni ore.

Marcasite, 42 mm, Vintířov, Czech Republic

Marcasite

FeS_2

ORTHORHOMBIC ● ● ● ●

Properties: C – tin-white, bronze-yellow, tarnishing iridescent; S – grayish to brownish- black; L – metallic; D – opaque; DE – 4,9; H – 6-6,5; CL – good; F – uneven; M – tabular, pyramidal and prismatic crystals, often twinned into the form of cockscomb-like aggregates, stalactitic, botryoidal and massive.

Origin and occurrence: It originates at low temperatures in very acidic environment, either in sedimentary, or in hydrothermal deposits, associated with pyrite, pyrrhotite, galena, sphalerite, fluorite, dolomite and calcite. Hydrothermal crystals and pseudo-morphs after pyrrhotite are known from Freiberg, Germany; Llallagua, Bolivia and Chiuzbaia, Romania. Crystals from Wiesloch, Germany and Reocin, Santander, Spain are of similar origin. Large crystals occurred in Joplin, Missouri and in Galena, Illinois, USA. The best crystals of sedimentary marcasite come from coal basins. Fine crystals from black coal are known from Essen, Germany. Cockscomb-like aggregates up to 150 mm (6 in) across, come from the brown coal basin in Vintírov, Czech Republic. Spherical, radial concretions with pyrite are known from Sparta, Illinois, USA and from Champagne, France.

Application: production of sulfuric acid.

Marcasite, 70 mm, Sparta, U.S.A.

Löllingite, 7 mm xx, Cobalt, Canada

Löllingite

$FeAs_2$

ORTHORHOMBIC ● ● ●

Properties: C – steel-gray to silver-white; S – gray-black; L – metallic; D – opaque; DE – 7,5; H – 5-5,5; CL – sometimes good; F – uneven; M – prismatic crystals, massive.
Origin and occurrence: Magmatic in pegmatites; hydrothermal in greisens and Sn-W veins, rare in the other types of ore veins, together with skutterudite, bismuth, nickeline, siderite, calcite and other minerals. Fine crystals are known from syenite in Langensundsfjord, Norway. Crystals up to 50 cm (20 in) across come from a pegmatite in Kaatiala, Finland. Masses occur in the Kobokobo pegmatite, Kivu, Zaire. Massive aggregates with schorl were found in Dolní Bory, aggregates with cassiterite in Prebuz, Czech Republic.

Safflorite

$(Co,Fe)As_2$

ORTHORHOMBIC ● ● ●

Properties: C – tin-white, tarnishing to dark gray; S – gray-black; L – metallic; D – opaque; DE – 7,5; H – 4,5-5; CL – good; F – uneven to conchoidal; M – prismatic crystals, radial aggregates, massive.
Origin and occurrence: Hydrothermal in medium-temperature with skutterudite, rammelsbergite, nickeline, silver, bismuth and löllingite. It is common in Schneeberg, Germany; Cobalt, Ontario, Canada; Batopilas, Chihuahua, Mexico; and in Bou Azzer, Morocco.

Rammelsbergite

$NiAs_2$

ORTHORHOMBIC ● ● ●

Properties: C – tin-white, pinkish; S – gray-black; L – metallic; D – opaque; DE – 7,1; H – 5,5-6; CL – good; F – uneven; M – prismatic crystals, radial and fibrous aggregates, massive.
Origin and occurrence: Hydrothermal in medium-temperature veins associated with other Ni and Co minerals. Botryoidal aggregates come from Schneeberg, Germany. It is common in Sarrabus, Sardinia, Italy; in the Eldorado mine near Great Bear Lake; in Cobalt, Ontario, Canada and Bou Azzer, Morocco.

Safflorite, 70 mm, Bou Azzer, Morocco

Rammelsbergite, 60 mm, Schneeberg, Germany

Arsenopyrite, 26 mm x, Portal, Mexico

Arsenopyrite

FeAsS

MONOCLINIC ● ● ● ●

Properties: C – silver-white to steel-gray; S – black; L – metallic; D – opaque; DE – 6,2; H – 5,5 – 6; CL – good; F – uneven; M – thick tabular to prismatic striated crystals, granular, massive.
Origin and occurrence: It occurs in pegmatites; hydrothermal in high-temperature vein deposits and greisens; metamorphic in contact metamorphic skarns, gneisses and mica schists. Long prismatic crystals up to 30 cm (12 in) long are known from the Obira mine, Japan. It is very common in greisen Sn and W deposits, fine crystals are known from Horní Slavkov, Czech Republic and Ehrenfriedersdorf, Germany. Beautiful crystals up to 50 mm (2 in) across come from Panasqueira, Portugal, where they occurred associated with fluorapatite, wolframite and siderite. Historically important were large crystals from Tavistock, Devon, UK. Crystals up to 40 mm (1⁹/₁₆ in) found in Llallagua, Bolivia. Crystals up to 50 mm (2 in) across were found recently in the Nikolai mine in Dalnegorsk, Russia.

Gudmundite, 10 mm xx, Polar Urals, Russia

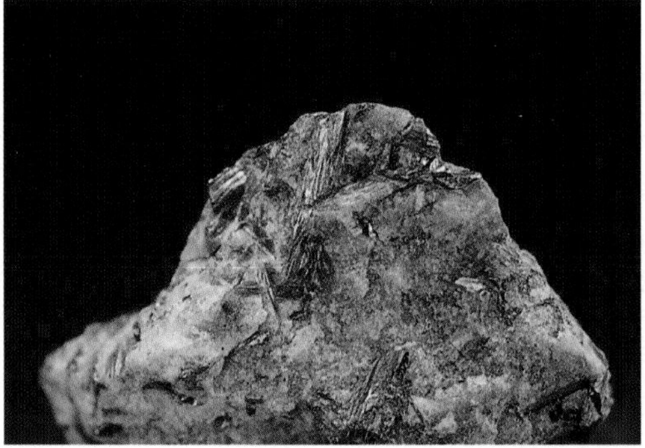

Shiny crystals up to 30 mm ($1^3/_{16}$ in) across, come from Hunan province, China. *Application:* As ore.

Gudmundite

FeSbS

MONOCLINIC ● ● ●

Properties: C – silver-white to steel-gray; S – black; L – metallic; D – opaque; DE – 7,0; H – 5-6; CL – none; F – uneven; M – prismatic twinned crystals, massive.
Origin and occurrence: Late hydrothermal mineral of ore deposits, also in metamorphic deposits and skarns. It is common in metamorphic sulfide deposits in Sweden (Boliden, Gudmundstorp). Massive aggregates are known from Kutná Hora and Vlastijovice, Czech Republic. It was common in Broken Hill, New South Wales, Australia.

Molybdenite

MoS_2

HEXAGONAL ● ● ● ●

Properties: C – lead-gray; S – blue-gray; L – metallic; D – opaque; DE – 4,0; H – 1-1,5; CL – perfect; M – tabular and prismatic crystals, scaly aggregates.
Origin and occurrence: Magmatic in pegmatites, granites and aplites, hydrothermal in high-temperature veins, also in porphyry ore deposits and in contact metamorphic deposits; associated with chalcopyrite, quartz and other minerals. Large crystals come from pegmatites in Blue Hill Bay, Maine, USA, and in Mutue-Fides-Stavoren, Transvaal, South Africa, where they reach several tens of cm in size. Crystals up to 150 x70 mm (6 x 2¾ in) across come from the transitional type between pegmatites and quartz veins near Arendal and Moss, Norway; large crystals also occur in the Temiskaming district, Quebec, Canada; tabular crystals up to 120 mm ($^{25}/_{32}$ in) across found in quartz-molybdenite breccia pipes in veins in Australia (Queensland, New South Wales). Fine crystals are also known from Kladnica near Vito_a, Bulgaria; Horní Slavkov, Czech Republic; and Ehrenfriedersdorf, Germany. As a fine grained disseminated ore was mined in Bingham, Utah and Climax, Colorado, USA.
Application: Mo ore.

Skutterudite

$CoAs_{2-3}$

CUBIC ● ● ●

Properties: C – tin-white to silver-gray, tarnishing to

Molybdenite, 26 mm, Molly Hill, U.S.A.

gray and iridescent; S – black; L – metallic; D – opaque; DE – 6,8; H – 5,5 – 6; CL – good; F – conchoidal to uneven; M – combinations of cubic crystals, skeletal aggregates, granular, massive.
Origin and occurrence: Hydrothermal in medium- to high-temperature ore veins, associated with other Ni and Co minerals. Crystals up to 50 mm (2 in) across come from Bou Azzer, Morocco. Crystals up to several cm in size were found in Schneeberg and Annaberg, Germany. Large massive accumulations occur in Cobalt and Gowganda, Ontario, Canada. *Application:* Co ore.

Skutterudite, 25 mm, Schneeberg, Germany

Nickel-skutterudite

$NiAs_3$

CUBIC ● ● ● ●

Properties: C – tin-white to silver, tarnishing to gray and iridescent; S – black; L – metallic; D – opaque; DE – 6,5; H – 5,5 – 6; CL – good; F – conchoidal to uneven; M – combinations of cubic crystals, skeletal aggregates, granular.
Origin and occurrence: Hydrothermal in medium-temperature veins with arsenopyrite, arsenic, bismuth, calcite and siderite. Known in crystals from Chatham, Connecticut and Chester, Massachusetts, USA; also Val d'Anniviérs, Wallis, Switzerland. Massive aggregates come from Dobšiná, Slovakia; Les Chalanches, France; Mohawk mine, Michigan, USA; Schneeberg, Germany. *Application:* Co and Ni ore.

Nickel-skutterudite, 36 mm, Saxony, Germany

Proustite, 25 mm, Chanarcillo, Chile

Proustite

Ag_3AsS_3

TRIGONAL ● ●

Properties: C – crimson, darkens upon exposure to light; S – crimson; L – adamantine; D – translucent to opaque; DE – 5,6; H – 2-2,5; CL – good; F – conchoidal to uneven; M – prismatic, rhombohedral and scalenohedral crystals, massive.
Origin and occurrence: Low-temperature hydrothermal mineral, also in the oxidation and cementation zone together with stephanite, silver, xanthoconite, acanthite and other minerals. The best specimens with crystals up to 100 mm (4 in) long come from the Dolores mine, Chañarcillo, Chile. Crystals up to 80 mm ($3^{1}/_{8}$ in) long found in the Himmelsfürst mine in Freiberg, Niederschlema and Schneeberg, Germany. Large druses with crystals up to 40 x 20 mm ($1^{9}/_{16}$ x $^{25}/_{32}$ in) across occurred in Jáchymov; crystals up to 20 mm ($^{25}/_{32}$ in) across known from Príbram and Stará Vožice, Czech Republic. Fine crystals come from Batopilas, Chihuahua and Sombrerete, Zacatecas, Mexico. Crystalline masses of proustite, weighing over 250 kg (550 lb), were found in 1865 in the Poorman mine, Silver City, Idaho, USA. *Application:* Ag ore.

Pyrargyrite, 61 mm, Zacatecas, Mexico

Pyrargyrite

Ag_3SbS_3

TRIGONAL ● ●

Properties: C – dark red, darkens upon exposure to light; S – crimson; L – adamantine; D – translucent to opaque; DE – 5,9; H – 2,5; CL – good; F – conchoidal to uneven; M – prismatic, rhombohedral and scalenohedral crystals, granular, massive.
Origin and occurrence: Low-temperature hydrothermal mineral, also secondary in the oxidation and cementation zone, together with silver, acanthite, other Ag sulfosalts, calcite and quartz. Crystals in Colquechaca, Bolivia and Chañarcillo, Chile reached several cm in size. Crystals up to 70 mm (2¾ in) long occurred in the Santo Niño vein in Fresnillo, Zacatecas, Mexico. San Genaro mine in Huancavelica, Peru yielded crystals up to 50 mm (2 in) across. Crystals up to 40 mm (1⁹/₁₆ in) across found in Freiberg; smaller crystals only are known from St. Andreasberg, Germany. Crystals up to 20 mm (²⁵/₃₂ in) across come from Příbram and Stará Vožice, Czech Republic. Crystals up to 50 mm (2 in) across found in the San Carlos mine, Hiendelaencina, Spain.
Application: Ag ore.

Xanthoconite

Ag_3AsS_3

MONOCLINIC ● ●

Properties: C – dark crimson, orange-yellow to yellow-brown; S – orange-yellow; L – adamantine; D – translucent; DE – 5,5; H – 2-3; CL – good; F – conchoidal; M – tabular and lath-like crystals, botryoidal and radial aggregates.

Xanthoconite, 2 mm xx, Marienberg, Germany

Pyrostilpnite, 60 mm, Potosí, Bolivia

Origin and occurrence: Hydrothermal in ore veins together with proustite, pyrargyrite, acanthite, arsenic and calcite. Botryoidal masses with yellow crystals up to 7mm (⁹/₃₂ in) long come from Freiberg; other important localities are St Andreasberg, Germany; Ste-Marie-aux- Mines, France; Cobalt, Ontario, Canada; Příbram, Třebsko; and Jáchymov, Czech Republic.

Pyrostilpnite

Ag_3SbS_3

MONOCLINIC ● ●

Properties: C – hyacinth- to orange-red; S – orange-yellow; L – adamantine; D – translucent; DE – 6,0; H – 2; CLB perfect; F – conchoidal; M – tabular to lath-like crystals, radial aggregates.
Origin and occurrence: Hydrothermal in low-temperature veins, associated with pyrargyrite, stephanite, acanthite and other Ag minerals. The best crystals come from St. Andreasberg, Germany. Crystals up to 10 mm (³/₈ in) long were found in Příbram, Třebsko and Jáchymov, Czech Republic. It is also described from Colquechaca, Bolivia and Chañarcillo, Chile.

Samsonite

$Ag_4MnSb_2S_6$

MONOCLINIC ●

Properties: C – steel-gray; S – dark red; L – metallic; D – opaque; DE – 5,5; H – 2,5; CL – none; F – conchoidal; M – prismatic striated crystals.
Origin and occurrence: The only locality, where it occurred in relatively larger amount, was the Samson mine in St. Andreasberg, Germany, where crystals up to 10 mm (3/8 in) across were found.

Samsonite, 12 mm x, St.Andreasberg, Germany

Chalcostibite

$CuSbS_2$

ORTHORHOMBIC ● ●

Properties: C – lead-gray, tarnishing to blue and green; S – lead-gray; L – metallic; D – opaque; DE – 5,0; H – 3-4; CL – perfect; F – conchoidal; M – long prismatic, striated crystals, granular, massive.
Origin and occurrence: Hydrothermal in ore veins, associated with jamesonite, chalcopyrite, tetrahedrite, stibnite, andorite and other minerals. Partly altered crystals up to 100 mm (4 in) long are known from Rar-el-Auz near Casablanca, Morocco. It is often in deposits in Bolivia (Huanchaca, Oruro, Colquechaca), where its crystals reach 10 mm (3/8 in)

Chalcostibite, 29 mm, Saint Pons, France

Emplectite, 60 mm, Schwarzenberg, Germany

in size. The world's best crystals up to 16 cm ($6^5/_{16}$ in) long were discovered recently near St Pons, France.

Emplectite

$CuBiS_2$

ORTHORHOMBIC ● ● ●

Properties: C – gray to tin-white; S – gray; L – metallic; D – opaque; DE – 6,4; H – 2; CL – perfect; F – conchoidal to uneven; M – thin prismatic to acicular striated crystals.
Origin and occurrence: Hydrothermal in high-temperature veins associated with chalcopyrite,

Berthierite, 98 mm, Herja, Romania

Wittichenite, 50 mm, Wittichen, Germany

molybdenite, quartz, tetrahedrite and other minerals. Fine acicular crystals up to 30 mm ($1^3/_{16}$ in) long come from Krupka, Czech Republic. Crystallized specimens were also found in Wittichen and Johanngeorgenstadt, Germany; Colquijirca, Peru; and in the Akenobe mine, Japan.

Wittichenite

Cu_3BiS_3

ORTHORHOMBIC ● ●

Properties: C – steel-gray to tin-white, tarnishing yellow to steel-gray; S – black; L – metallic; D – opaque; DE – 6,2; H – 2-3; CL – none; F – conchoidal; M – prismatic crystals, massive.
Origin and occurrence: Hydrothermal in ore veins, associated with other Bi minerals, Cu-Fe sulfides, selenides and secondary U minerals. It occurs in Wittichen, Germany; Baiţa Bihorului, Romania; Tsumeb, Namibia; and Cerro de Pasco, Peru.

Berthierite

$FeSb_2S_4$

ORTHORHOMBIC ● ● ●

Properties: C – dark steel-gray, tarnishing iridescent to brown; S – dark brown-gray; L – metallic; D – opaque; DE – 4,7; H – 2-3; CL – imperfect; M – long prismatic, striated crystals, fibrous, felt-like and radial aggregates.
Origin and occurrence: Hydrothermal in low-temperature Sb deposits. Acicular crystals up to 10 mm ($^3/_8$ in) long are known from the St. Antoni de Padua gallery in Kutná Hora, Czech Republic; thick prismatic crystals come from Poproè, Slovakia. Iridescent columnar aggregates up to 200 mm ($7^7/_8$ in) long occur in Herja, Romania. Fine specimens are known also from Oruro, Bolivia.

Stephanite, 8 mm xx, Zacatecas, Mexico

Stephanite

Ag_5SbS_4

ORTHORHOMBIC ● ●

Properties: C – black; S – black; L – metallic; D – opaque; DE – 6,3; H – 2-2,5; CL – imperfect; F – conchoidal; M – short prismatic to tabular striated crystals, massive.
Origin and occurrence: Late hydrothermal mineral in Ag deposits, associated with proustite, acanthite, silver, tetrahedrite, galena, sphalerite and pyrite. Crystals up to 40 mm ($1^{9}/_{16}$ in) long come from Příbram and Jáchymov, Czech Republic and from St. Andreasberg, Germany. Crystals up to 50 mm (2 in) across found in the Las Chispas mine, Arizpe, Sonora, Mexico and Hiendelaencina, Spain. Smaller crystals occurred in Freiberg, Schneeberg and Annaberg, Germany. *Application:* Ag ore.

Pearceite

$Ag_{16}As_2S_{11}$

MONOCLINIC ● ●

Properties: C – black; S – black; L – metallic; D – opaque; DE – 6,1; H – 3; CL – none; F – conchoidal to irregular; M – short prismatic to tabular crystals and rosette-like aggregates, massive.
Origin and occurrence: Hydrothermal in low- to medium-temperature deposits with acanthite, silver, proustite, quartz, barite and calcite. Crystals several mm across are known from Jáchymov, Moldava and Medenec, Czech Republic; from Arqueros, Chile and from the Veta Rica mine, Coahuila, Mexico. Crystals up to 12 mm ($^{15}/_{32}$ in) across were found in the Caribou mine, Colorado, USA, and in Dzhezkazgan, Kazakhstan. Huge accumulations of almost pure pearceite occurred in the Mollie Gibson mine near Aspen, Colorado, USA.
Application: Ag ore.

Polybasite

$(Ag,Cu)_{16}Sb_2S_{11}$

MONOCLINIC ● ●

Properties: C – black; S – black; L – metallic; D Bopaque; DE – 6,4; H – 2-3; CL – imperfect; F – uneven; M – pseudo-hexagonal tabular crystals, massive.
Origin and occurrence: Hydrothermal in low- to medium-temperature ore veins, associated with pyrargyrite, tetrahedrite, stephanite, acanthite,

Pearceite, 15 mm xx, Guanajuato, Mexico

Polybasite, 39 mm, Zacatecas, Mexico

quartz and other minerals. Crystals, several cm across, are known from Wolfach, St. Andreasberg, Freiberg and Schneeberg, Germany; also from Guanajuato, Mexico. The best specimens with tabular crystals up to 90 mm ($3^9/_{16}$ in) across come from the Las Chispas mine, Arizpe, Sonora, Mexico.
Application: Ag ore.

Lorandite
$TlAsS_2$

MONOCLINIC ● ●

Properties: C – crimson, lead-gray, it covers with a yellow coating; S – cherry-red; L – metallic to adamantine; D – translucent to transparent; DE – 5,5; H – 2-2,5; CL – perfect; M – short prismatic to tabular crystals, granular, massive.
Origin and occurrence: Hydrothermal, associated with stibnite, realgar, orpiment, pyrite and other minerals. Crystals up to 50 mm (2 in) across were found in Allchar, Macedonia. It is also known from Djijikrut, Tajikistan and from the cavities in dolomite from Lengenbach, Binntal, Switzerland.

Livingstonite
$HgSb_4S_8$

MONOCLINIC ● ●

Properties: C – black-gray; S – red; L – metallic to adamantine; D – opaque; DE – 5,0; H – 2; CL – perfect; M – acicular crystals, columnar and fibrous aggregates, massive.
Origin and occurrence: Hydrothermal in low-temperature veins, associated with cinnabar, stibnite, getchellite and other minerals. The best specimens with prismatic crystals up to 50 mm (2 in) long are known from Khaidarkan, smaller crystals only found from Kadamdzhai, Kyrgyzstan. It is also described from the La Cruz mine, Huitzuco, Guerrero, Mexico and from the Matsuo mine, Japan.

Lorandite, 10 mm x, Allchar, Macedonia

Livingstonite, 40 mm, Khaidarkan, Kyrgystan

Bournonite, 62 mm, Cornwall, UK

Bournonite
$PbCuSbS_3$

ORTHORHOMBIC ● ● ● ●

Properties: C – steel-gray to black; S – steel-gray to black; L – adamantine to dull; D – opaque; DE – 5,8; H – 2,5-3; CL – imperfect; F – conchoidal to uneven; M – short prismatic to tabular crystals, often striated, granular, massive.
Origin and occurrence: Hydrothermal in medium-temperature ore deposits, together with galena, tetrahedrite, pyrite, siderite and other minerals. The finest crystals, complex twins called cogswheel ore, over 50 mm (2 in) across come from the Herodsfoot mine near Liskeard, Cornwall, UK. Tabular crystals up to 40 mm ($1^{9}/_{16}$ in) across found in siderite cavities in Příbram, Czech Republic; large prismatic and tabular crystals known from Neudorf, Germany. Smaller crystals common in Cavnic and Baia Sprie, Romania. Crystals, up to 100 mm (4 in) across, occurred in Machacamarca, Bolivia. Crystals up to 40 mm ($1^{9}/_{16}$ in) come from Huancavelica, Peru. Crystals up to 100 mm (4 in) across reported from the Les Malines mine, France. Crystals up to 20 mm (25/32 in) across found at Chenzhou, Hunan province, China. *Application:* Pb, Cu and Sb ore.

Aikinite
$PbCuBiS_3$

ORTHORHOMBIC ● ●

Properties: C – black-gray, tarnishing brown; S – gray-black; L – metallic; D – opaque; DE – 7,3; H –

Aikinite, 2 mm xx, Rudnany, Slovakia

Betekhtinite, 10 mm xx, Dzhezkazgan, Kazakhstan

2-2,5; CL – imperfect; F – uneven; M – prismatic to acicular, striated crystals, massive.
Origin and occurrence: Hydrothermal in ore veins, associated with gold, pyrite, galena, tennantite and other minerals. It is common in quartz veins with gold in Berezovsk, Ural Mountains, Russia, where crystals up to 30 mm (1 3/16 in) long were found; fine crystals come also from La Gardette, Bourg d'Oisans, France. Grains up to 50 mm (2 in) across occurred in the Outlaw mine, Nevada, USA. It is also known in metamorphic veins in Val d'Anniviers, Switzerland.

Betekhtinite
$Cu_{10}(Fe,Pb)S_6$

ORTHORHOMBIC ● ●

Properties: C – black; S – black; L – metallic; D – opaque; DE – 6,1; H – 3-3,5; CL – good; M – acicular crystals, granular.
Origin and occurrence: Hydrothermal in ore deposits. The best specimens come from Dzhezkazgan as clusters of acicular crystals up to 70 mm (2¾ in) long, associated with bornite, chalcocite, djurleite and other minerals. Rich specimens found in Kipushi Kipushi, Shaba, Zaire. Granular aggregates fairly common in calcite veinlets, cross-cutting Cu-bearing shales near Eisleben, Mansfeld, Germany.

Andorite
$PbAgSb_3S_6$

ORTHORHOMBIC ● ●

Properties: C – dark steel-gray, tarnishing to yellow and iridescent; S – black; L – metallic; D – opaque; DE – 5,4; H – 3-3,5; CL – none; F – conchoidal; M – prismatic and tabular striated crystals, massive.
Origin and occurrence: Hydrothermal in ore deposits, together with cassiterite, jamesonite, stannite and other minerals. The world's best specimens come from the Itos and San José mines in Oruro and the Tatasi mine in Potosí, Bolivia, where it forms crystals up to 30 mm (1 3/16 in) across. Thin tabular crystals are known from Baia Sprie, Romania and from the Keyser mine, Nevada, USA. Needles up to 10 mm (3/8 in) long occurred in Třebsko, Czech Republic.
Application: ruda Ag.

Andorite, 45 mm, Oruro, Bolivia

Freieslebenite, 2 mm xx, Baia Sprie, Romania

Freieslebenite
$AgPbSbS_3$

MONOCLINIC ● ●

Properties: C – light steel-gray, lead-gray to silver-white; S – light steel-gray, lead-gray to silver-white; L – metallic; D – opaque; DE – 6,2; H – 2-2,5; CL – imperfect; F – conchoidal to uneven; M – prismatic, striated crystals.
Origin and occurrence: Hydrothermal in ore deposits, associated with acanthite, pyrargyrite, silver, galena, siderite and andorite. Crystals up to 20 mm ($^{25}/_{32}$ in) across come from the Santa Cecilia, Guadalajara, Spain; Freiberg, Germany; Oruro, Bolivia; and from the Treasury Lode, Colorado, USA.

Diaphorite
$Pb_2Ag_3Sb_3S_3$

MONOCLINIC ● ●

Properties: C – steel-gray; S – steel-gray; L – metallic; D – opaque; DE – 6,0; H – 2,5-3; CL – none; F – conchoidal to uneven; M – prismatic striated crystals.
Origin and occurrence: Hydrothermal in medium-temperature ore veins, associated with galena, sphalerite, pyrargyrite, pyrite and other minerals.

Diaphorite, 5 mm xx, Příbram, Czech Republic

Sartorite, 4 mm x, Binntal, Switzerland

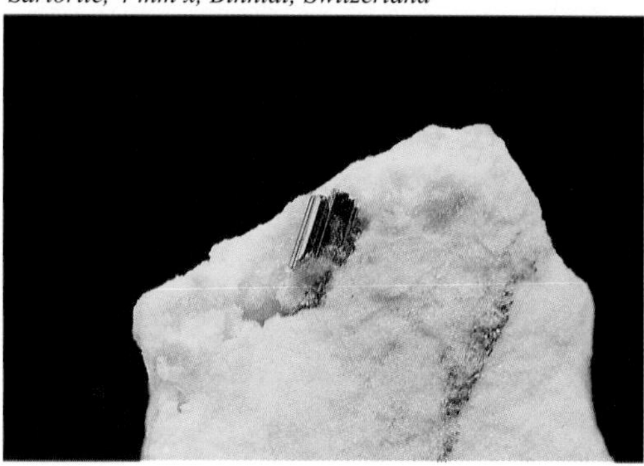

Beautiful striated crystals up to 10 mm (3/8 in) come from the cavities of quartz veins in Príbram; rare small crystals, several mm across found in the cavities of quartz veins in the St. Antoni de Padua gallery in Kutná Hora, Czech Republic. Complicated combinations of crystals were described from Bräunsdorf, Germany. Crystals up to 80 mm (3 1/8 in) long occurred in Hiendelaencina, Guadalajara, Spain. It is also reported from Catorce, San Luis Potosí, Mexico.

Sartorite

$PbAs_2S_4$

MONOCLINIC ● ●

Properties: C – dark lead-gray; S – chocolate-brown; L – metallic; D – opaque; DE – 5,1; H – 3; CL – good; F – conchoidal; M – prismatic striated crystals.
Origin and occurrence: Hydrothermal in dolomite, associated with tennantite, dufrenoysite, pyrite and realgar. The best crystals up to 100 mm (4 in) long come from Lengenbach, Binntal, Switzerland. It was also found in the Zuni mine, Colorado, USA.

Baumhauerite

$Pb_3As_4S_9$

TRICLINIC ●

Properties: C – lead- to steel-gray, tarnishing iridescent; S – chocolate-brown; L – metallic; D – opaque; DE – 5,4; H – 3; CL – perfect; F – conchoidal; M – tabular to short prismatic striated crystals, granular.
Origin and occurrence: Hydrothermal, associated with realgar and other sulfosalts. The best crystals up to 25 mm (1 in) across come from Lengenbach, Binntal, Switzerland. Massive aggregates were found in Hemlo, Thunder Bay, Ontario, Canada and Sterling Hill, New Jersey, USA.

Baumhauerite, 50 mm, Binntal, Switzerland

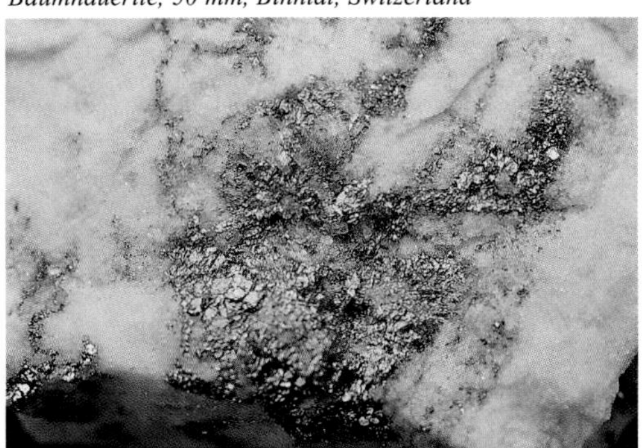

Rathite, 60 mm, Binntal, Switzerland

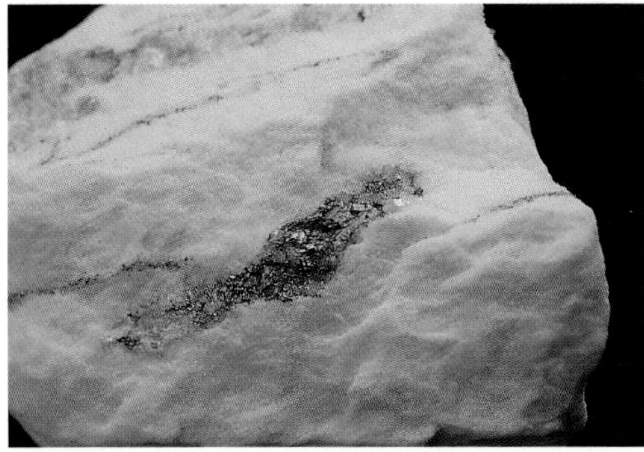

Rathite

$(Pb,Tl)_3As_5S_{10}$

MONOCLINIC ●

Properties: C – lead-gray, tarnishing iridescent; S – chocolate-brown; L – metallic; D – opaque; DE – 5,3; H – 3; CL – perfect; F – conchoidal; M – prismatic striated crystals.
Origin and occurrence: Hydrothermal, associated with other Pb-Tl-As-S minerals. Crystals up to 10 mm (3/8 in) across come from dolomite in Lengenbach, Binntal, Switzerland.

Dufrenoysite

$Pb_2As_2S_5$

MONOCLINIC ● ●

Properties: : C – lead- to steel-gray; S – red-brown to chocolate-brown; L – metallic; D – opaque; DE – 5,6; H – 3; CL – perfect; F – conchoidal; M – elongated striated tabular crystals.
Origin and occurrence: Hydrothermal low-temperature mineral, associated with rathite, sartorite, baumhauerite and realgar. Crystals up to 25 mm (1 in) across come from dolomite in Lengenbach, Binntal, Switzerland. It is also found in Batopilas, Chihuahua, Mexico.

Dufrénoysite, 10 mm x, Binntal, Switzerland

Jordanite, 12 mm x, Binntal, Switzerland

Jordanite
$Pb_{14}(As,Sb)_6S_{23}$

MONOCLINIC ● ●

Properties: C – lead-gray, tarnishing iridescent; S – black; L – metallic; D – opaque; DE – 6,4; H – 3; CL – perfect; F – conchoidal; M – tabular crystals, botryoidal aggregates.

Origin and occurrence: Hydrothermal in low-temperature ore veins, also in metamorphic dolomites, together with tennantite, sphalerite, galena, dolomite and other minerals. The most famous crystals up to 50 mm (2 in) across come from cavities in dolomite in Lengenbach, Binntal, Switzerland. Tabular crystals occurred also in Săcărîmb, Romania. Botryoidal aggregates, growing on barite crystals, were described from the Yunosawa mine, Japan.

Geocronite, 70 mm, Příbram, Czech Republic

Geocronite
$Pb_{14}(Sb,As)_6S_{23}$

MONOCLINIC ● ●

Properties: C – light lead-gray; S – light lead-gray to gray-blue; L – metallic; D – opaque; DE – 6,4; H – 2,5; CL – good; F – uneven; M – tabular crystals, massive.
Origin and occurrence: Hydrothermal in ore veins, associated with galena, pyrite, tetrahedrite, barite, fluorite and quartz. Crystals, up to 80 mm ($3^{1}/_{8}$ in) across, come from Pietrasanta, Italy. Crystals up to 90 mm ($3^{9}/_{16}$ in) found in Virgem da Lapa, Brazil. Tabular crystals up to 40 mm ($1^{9}/_{16}$ in) across occurred in the Kilbricken mine, Ireland. Massive aggregates are known from Príbram, Czech Republic.

Zinkenite

$Pb_9Sb_{22}S_{42}$

HEXAGONAL ● ● ●

Properties: C – steel-gray, tarnishing iridescent; S – steel-gray; L – metallic; D – opaque; DE – 5,3; H – 3-3,5; CL – imperfect; F – uneven; M – thin prismatic striated crystals, radial to felt-like aggregates, massive.

Origin and occurrence: Hydrothermal in ore veins, associated with stibnite, jamesonite, boulangerite, bournonite, stannite and other minerals. Crystals up to 50 mm (2 in) across are known from the Itos and San José mines in Oruro, Bolivia. It was also found in Wolfsberg, Germany; St. Pons, France; Săcărîmb; and Baia Sprie, Romania.

Jamesonite

$Pb_4FeSb_6S_{14}$

MONOCLINIC ● ● ● ●

Properties: C – gray-black, tarnishing iridescent; S – gray-black; L – metallic; D – opaque; DE – 5,8; H – 2,5; CL – good; F – uneven; M – acicular crystals, fibrous and felt-like aggregates, massive.

Zinkenite, 39 mm, Oruro, Bolivia

Jamesonite, 30 mm xx, Sombrerete, Mexico

Origin and occurrence: Hydrothermal in medium- and low-temperature base metal ore veins, associated with other Pb-sulfosalts, pyrite, sphalerite, galena, tetrahedrite, quartz and other minerals. Needles up to 80 mm ($3^1/_8$ in) long occur in many localities in Bolivia (Tasna; Bolivia mine, Poopó; San José and Itos mines, Oruro). It also comes from Wolfsberg and Freiberg, Germany; Nižná Slaná, Slovakia; and Sombrerete, Zacatecas, Mexico.
Application: Pb and Sb ore.

Semseyite

$Pb_9Sb_8S_{21}$

MONOCLINIC ● ●

Properties: C – gray to black; S – black; L – metallic; D – opaque; DE – 6,1; H – 2,5; CL – perfect; M – tabular and prismatic crystals and their rosette-like aggregates.
Origin and occurrence: Hydrothermal in medium-temperature veins, associated with jamesonite, bournonite, zinkenite, sphalerite and other minerals. Fine rosette-like aggregates of crystals over 10 mm (3/8 in) across come from Baia Sprie, Herja and Rodna, Romania. It was also found in the San José mine, Oruro, Bolivia; Wolfsberg, Germany; and Huancavelica, Peru.

Boulangerite

$Pb_5Sb_4S_{11}$

MONOCLINIC ● ● ● ●

Properties: C – lead-gray; S – brownish; L – metallic to silky; D – opaque; DE – 6,2; H – 2,5-3; CL – good; M – acicular striated crystals, fibrous and felt-like aggregates.
Origin and occurrence: Hydrothermal in low- and medium-temperature ore veins, together with other Pb sulfosalts, galena, sphalerite and other minerals. Fine needles over 100 mm (4 in) long come from cavities in quartz in Příbram, Czech Republic. It is common in the Coeur d'Alene district, Idaho, USA. Acicular crystals up to 30 cm (12 in) long were found in Trepèa, Serbia and Leadville, Colorado, USA. It is also known from Wolfsberg, Germany and Bolivia (Colquechaca, Huanuni, Isca-Isca).

Cosalite

$Pb_2Bi_2S_5$

ORTHORHOMBIC ● ● ●

Properties: C – lead- to steel-gray, silver-white; S – black; L – metallic; D – opaque; DE – 7,1; H – 2,5-3; CL – none; F – uneven; M – prismatic to acicular crystals, radial and fibrous aggregates.
Origin and occurrence: Magmatic in pegmatites, hydrothermal in medium-temperature deposits; also metamorphic, associated with sphalerite, chalco-

Semseyite, 40 mm, Cavnic, Romania

Boulangerite, 40 mm, Zacatecas, Mexico

Cosalite, 3 mm xx, Ocna de Fier, Romania

pyrite, pyrite, cobaltite and other minerals. Elongated crystals are known from Crodo, Italy. Fine needles up to 40 mm ($1^{9}/_{16}$ in) long, included in quartz crystals, were found in Kara-Oba, Kazakhstan. It occurs also in Au deposits (Homestake mine, South Dakota, USA) or in skarns (Baiţa Bihorului, Romania).

Kobellite

$Pb_{22}Cu_4(Bi,Sb)_{30}S_{69}$

ORTHORHOMBIC ● ● ●

Properties: C – black-gray to steel-gray; S – black; L – metallic; D – opaque; DE – 6,5; H – 2,5-3; CL – good; M – fibrous aggregates, granular, massive.
Origin and occurrence: Hydrothermal in high-temperature veins and pegmatites, together with cobaltite, arsenopyrite, chalcopyrite and other minerals. It is known from a sulfide rich pegmatite in the Superior Stone quarry, North Carolina, USA. Massive aggregates are common in Jedl'ovec, Slovakia. It was originally described from the Hvena mine near Åskersund, Sweden.

Kobellite, 20 mm xx, Rožňava, Slovakia

Realgar, 64 mm, Shimen, China

Realgar

AsS

MONOCLINIC ● ● ● ●

Properties: C – red to orange-yellow; S – orange-red to red; L – resinous to greasy; D – transparent to translucent; DE – 3,6; H – 1,5-2; CL – good; F – conchoidal; M – prismatic striated crystals, granular, massive.

Orpiment, 40 mm, Baia Sprie, Romania

Origin and occurrence: Hydrothermal in low-temperature veins, associated with other As-Sb minerals; also as a sublimation product of volcanic gasses, in hot springs and sediments. The most beautiful crystals over 100 mm (4 in) long come from Shimen, Hunan, China. Crystals up to 70 mm ($2\frac{3}{4}$ in) long occurred in the Getchell mine, Nevada, USA. Crystals up to 80 mm ($3\frac{1}{8}$ in) long were found in Lengenbach, Binntal, Switzerland. Crystals are also known from Baia Sprie, Romania. Massive aggregates are common in Allchar, Macedonia.

Orpiment

As_2S_3

MONOCLINIC ● ● ● ●

Properties: C – lemon-yellow to bronze-yellow; S – light lemon-yellow; L – resinous to pearly; D – transparent to translucent; DE – 3,5; H – 1,5-2; CL – perfect; M – prismatic crystals, foliated and fibrous aggregates. *Origin and occurrence:* Hydrothermal in low-temperature veins, together with realgar, stibnite, calcite etc., also from hot springs and fumaroles. It is also a common product of realgar oxidation. The best crystals up to 100 mm (4 in) long come from Shimen, Hunan, China. Fine cleavable lamellae occur in Lukhumi, Georgia and Men-Kyule, Yakutia, Russia. Crystals up to 50 mm (2 in) across found in the La Libertad mine, Quiruvilca and Huayllapon, Ancash, Peru. Crystals up to 80 mm (3¹/₈ in) long described from the Getchell mine, Nevada, USA. Fine specimens are also known from Allchar, Macedonia and Khaidarkan, Kyrgyzstan. *Application:* As ore, pigment.

Getchellite

$AsSbS_3$

MONOCLINIC ● ●

Properties: C – dark red, tarnishing green and iridescent; S – orange-red; L – pearly to glassy, resinous; D – transparent; DE – 4,0; H – 1,5-2; CL – perfect; F – splintery; M – imperfect curved crystals, massive.
Origin and occurrence: Hydrothermal in low-temperature ore deposits, associated with orpiment, realgar, stibnite, cinnabar and other minerals. It was described from the Getchell mine, Nevada, USA. Beautiful specimens with grains up to several cm across come from Khaidarkan, Kyrgyzstan. It is also known from Zarehshuran, Kurdistan, Iran.

Getchellite, 60 mm, Khaidarkan, Kyrgystan

Orpiment, 35 mm, Huayllapon, Peru

3. Halides

Fluellite

$Al_2 (PO_4)F_2(OH) . 7 H_2O$

ORTHORHOMBIC ● ●

Properties: C – colorless, white, yellow; S – white; L – vitreous; D – transparent; DE – 2.2; H – 3; CL – imperfect; M – dipyramidal crystals.
Origin and occurrence: Hydrothermal in greisens, also secondary as a result of triplite alteration. Crystal druses over 10 mm (3/8 in) in size come from Horní Slavkov, Czech Republic. Very similar specimens were found in Stenna Gwyn near St. Austell, Cornwall, UK. Also found in pegmatites in Kynzvart, Czech Republic and in Hagendorf, Germany as a product of triplite alteration.

Fluellite, 10 mm x, Horní Slavkov, Czech Republic

Cryolite

Na_3AlF_6

MONOCLINIC ● ● ●

Properties: C – colorless, white, purple, brownish; S – white; L – greasy to pearly; D – transparent to translucent; DE – 3; H – 2.5; CL – none; F – uneven; M – pseudo-cubic crystals, massive.

Origin and occurrence: Characteristic mineral of the cryolite pegmatites. Crystals up to 30 mm (1 3/16 in) in size were found in Ivigtut, Greenland, where it was mined as the Al ore for more than 100 years. It was associated with other aluminofluorides, sphalerite, cassiterite, ferrocolumbite and other minerals. It is also known from the Francon quarry in Montreal, Quebec, Canada in crystals up to 10 mm (3/8 in) across. Massive cryolite occurs in Miass, Ural Mountains, Russia and in St. Peter's Dome, Pikes Peak batholith, Colorado, USA.
Application: it was an important Al ore.

Fluorite, 67 mm, Berbes, Spain
Cryolite, 35 mm, Ivigtut, Greenland

Creedite, 70 mm, Santa Eulalia, Mexico

Creedite

$Ca_3Al_2(SO_4)(F,OH)_{10} . 6\ H_2O$

ORTHORHOMBIC ●●●

Properties: C – colorless, white, purple; S – white; L – vitreous; D – transparent; DE – 2.7; H – 4; CL – perfect; F – conchoidal; M – short prismatic to acicular crystals, granular, massive.
Origin and occurrence: Hydrothermal, associated with fluorite and barite. Purple crystals, several cm long, come from Wagon Wheel Gap near Creede, Colorado, USA. Nice druses were found in Santa Eulalia, Chihuahua, Mexico. The best creedite specimens with purple crystals up to 30 mm (1 3/16 in) long were recently found in Akcha-tau, Kazakhstan.

Carnallite

$KMgCl_3 . 6\ H_2O$

ORTHORHOMBIC ●●●●

Properties: C – colorless, white, yellowish, red, blue; S – white; L – vitreous to greasy; DE – 1.6; H – 1-2; CL – none; F – conchoidal; M – pseudo-hexagonal pyramidal and tabular crystals, granular; LU – strong; R – decomposes under wet conditions.
Origin and occurrence: Sedimentary, one of the last products of evaporation of salty solutions; also supergene as a product of a reaction of older salts with solutions, rich in potassium, associated with halite, sylvite and other minerals. Crystals up to 40 mm (1 9/16 in) across are known from the vicinity of Carlsbad, New Mexico, USA. Nice crystals come also from Stassfurt and Alexanderhall, Germany. Massive aggregates are common in many salt deposits, like in Saskatchewan, Canada; in Kalush, Ukraine and elsewhere.
*Application:*the most important potassium salt, used as fertilizer and for production of metal Mg.

Atacamite

$Cu_2Cl(OH)_3$

ORTHORHOMBIC ●●●

Properties: C – emerald-green, black-green; S – green; L – vitreous; D – translucent; DE – 3.8; H – 3-3.5; CL – perfect; F – conchoidal; M – prismatic crystals, columnar, radial and lamellar aggregates, granular, massive.
Origin and occurrence: Secondary in the oxidation zone of Cu deposits in the arid climate, associated with other Cu minerals. Crystals up to 10 mm (3/8 in) long were described from Burra district, Southern Australia, Australia. Crystals up to 10 mm (3/8 in) across come from Tsumeb, Namibia; also from Bisbee, Arizona, USA. Rich aggregates of acicular crystals occur in many localities in Atacama province, Chile (Copiapó, Remolinos).

Carnallite, 80 mm, Merkers, Germany

Atacamite, 78 mm, Atacama, Chile

Boleite

$Pb_{26}Ag_9Cu_{24}Cl_{62}(OH)_{48}$

CUBIC ●

Properties: C – blue; S – blue; L – pearly; D – translucent; DE – 5.1; H – 3-3.5; CL – perfect; M – cubic crystals; R – soluble in water.

Origin and occurrence: Secondary, originated in the oxidation zone of Cu deposits in the arid climate. By far the best specimens were found in Boléo, Baja California, Mexico, where cubes up to 25 mm (1 in) in size were found. It is also known from Phillipsburg, Montana, USA and Challacollo, Chile.

Boleite, 65 mm, Santa Rosalia, Mexico

Cumengite, 21 mm, Boléo, Mexico

Cumengite

$Pb_{21}Cu_{20}Cl_{42}(OH)_{40}$

TETRAGONAL ●

Properties: C – indigo-blue; S – blue; L – vitreous; D – translucent; DE – 4.7; H – 2.5; CL – good; M – tetragonal pyramidal crystals, also epitaxially overgrown on boleite cubes. *Origin and occurrence:* Secondary in the oxidation zone of Cu deposits in the arid climate, associated with boleite. The largest crystals up to 35 mm (1 3/8 in) are known from Boléo, Baja California, Mexico. It is also reported from Newport Beach near Falmouth, Cornwall, UK.

Iodargyrite

β -AgI

HEXAGONAL ● ● ●

Properties: C – colorless, tarnishes to yellow; S – yellow; L – adamantine; D – transparent to translucent; DE – 5.7; H – 1-1.5; CL – perfect; F – conchoidal; M – prismatic to tabular crystals, granular, massive.

Origin and occurrence: Secondary, product by oxidation of Ag ores, with other Ag minerals. Common greenish crystals over 10 mm (3/8 in) in size occur in Broken Hill, New South Wales, Australia. Also found in Vrancice, Czech Republic; Tonopah, Nevada, USA; Chañarcillo and Copiapó, Chile.

Villiaumite

NaF

CUBIC ● ● ●

Properties: C – dark red; S – white; L – vitreous; D – transparent to translucent; DE – 2.8; H – 2-2.5; CL – perfect; M – small crystals, granular, massive; R – soluble in water.

Origin and occurrence: Late mineral in cavities in alkaline igneous rocks (nepheline syenites). Crystals, several cm long, are known from the Rasvumchorr Mountain, Khibiny massif, Kola Peninsula, Russia; only slightly smaller crystals come from Mont St.-Hilaire, Quebec, Canada and Illimaussaq, Greenland.

Iodargyrite, 2 mm xx, Rudabánya, Hungary

Villiaumite, 10 mm xx, Khibiny Massif, Kola, Russia

Halite, 55 mm, Trona, U.S.A.

Halite
NaCl

CUBIC ● ● ● ●

Properties: C – colorless, gray, white, red, blue; S – white; L – vitreous; DE – 2.2; H – 2; CL – perfect; F – conchoidal; M – cubic crystals, granular, massive; R – soluble in water.
Origin and occurrence: Product of high-temperature fumaroles (Etna, Mt. Vesuvius; Italy); mainly sedimentary, as a result of evaporation of sea water, associated with sylvite, carnallite and other minerals. Very fine cubes over 10 mm (3/8 in) are known from Weliczka and Bochnia, Poland. Blue cleavable aggregates are found in Bernburg, Germany. Salt deposits in Austria (Hallstatt, Hallein) are also important. Huge halite deposits, associated with potassium salts are mined in the vicinity of Stassfurt, Germany. Fine skeletal crystals are known from many localities in California, USA.
Application: food and chemical industries.

Halite, 18 mm xx, Sonora, Mexico

Sylvite, 77 mm, California, U.S.A.

Sylvite
KCl

CUBIC ● ● ●

Properties: C – white, gray, blue, red; S – white; L – vitreous; D – transparent; DE – 2; H – 2; CL – perfect; F – uneven; M – cubic crystals and their combinations; granular, massive; R – soluble in water.
Origin and occurrence: Sedimentary as a result of evaporation of sea water, together with halite, carnallite and other minerals. Nice cubes up to 50 mm (2 in) come from Stassfurt; it forms stalactites in Wathlingen, Germany. Crystals are also known from Kalush, Ukraine and from Salton Sea, California, USA, where it occurs as octahedra on halite crystals.
Application: chemical industry.

Sal ammoniac, 10 mm x, Bátonyterenye, Hungary

Sal ammoniac
NH_4Cl

CUBIC ● ● ●

Properties: C – colorless, white, gray, yellow, brown; S – white; L – vitreous; D – transparent; DE – 1.5; H – 1-2; CL – imperfect; F – conchoidal; M – combinations of cubic crystals, dendritic and skeletal aggregates; earthy.
Origin and occurrence: Typical mineral for fumaroles and burning coal dumps, associated with sulfur and other minerals. Complicated crystals are known from Mt. Vesuvius, Etna and Vulcano, Italy. Crystals over 10 mm ($^3/_8$ in) in size occurred on burning coal dumps near Kladno, Czech Republic; similar from localities in Eastern Pennsylvania, USA and near Ste-Etienne, France.

Calomel, 30 mm, Terlingua, U.S.A.

Calomel
HgCl

TETRAGONAL ● ●

Properties: C – colorless, white, gray, brown, it darkens on air; S – white; L – adamantine; D – transparent to translucent; DE – 7.2; H – 1.5; CL – good; F – conchoidal; M – tabular to pyramidal crystals, coatings, earthy; LU – dark red.
Origin and occurrence: Secondary as a result of alteration of Hg minerals, associated with cinnabar, mercury and other minerals. Crystals were found in Moschellandsberg, Germany; Avala, Serbia;

Khaidarkan, Kyrgyzstan; and Terlingua, New Mexico, USA.

Fluorite

CaF_2

CUBIC ● ● ● ● ●

Properties: C – colorless, white, yellow, red, green, blue, purple, brown, black; S – colorless; L – vitreous; D – transparent to translucent, opaque; DE – 3.2; H – 4; CL – perfect; F – conchoidal to splintery; M – combinations of cubic crystals, granular, massive; LU – blue, blue-green, also phosphorescent.
Origin and occurrence: Rare magmatic, mainly hydrothermal and metasomatic. Associations are very diverse, depending on a type of the deposit, in which it occurs. Beautiful crystals are known from many localities all over the world. Pink octahedra, several cm in size, are known from pegmatites in Nagar, Pakistan. Nice crystals were also found in greisens in Cornwall, UK (Wheal Mary mine) and from Horní Slavkov, Czech Republic. Beautiful green cubes up to 20 cm across and colorless cubes up to 10 mm (3/8 in) across from Dalnegorsk, Russia are of hydrothermal origin. Famous green and purple crystals come from Alston Moor and Weardale, England, UK. Nice pink octahedra up to 30 mm (1 3/16 in) occurred in Huanzala, Peru. Beautiful yellow cubes up to 50 mm (2 in), associated with barite, are known from Halsbrücke and Annaberg, Germany. Purple complex combinations of crystals come from La Collada, Spain. Mainly purple cubes up to 10 mm (3/8 in) occurred in Rosiclare, Illinois; in association with honey-yellow calcite crystals are known from the Elmwood mine, Tennessee; similar occurrences are also in several localities in Kentucky, USA. The most valuable fluorite specimens are pink octahedra up to 150 mm (6 in) from Göschenen, Switzerland; Mont Blanc massif, France; and other Alpine localities.
Application: metallurgy, chemical industry, special optics.

Fluorite, 50 mm, Argentina

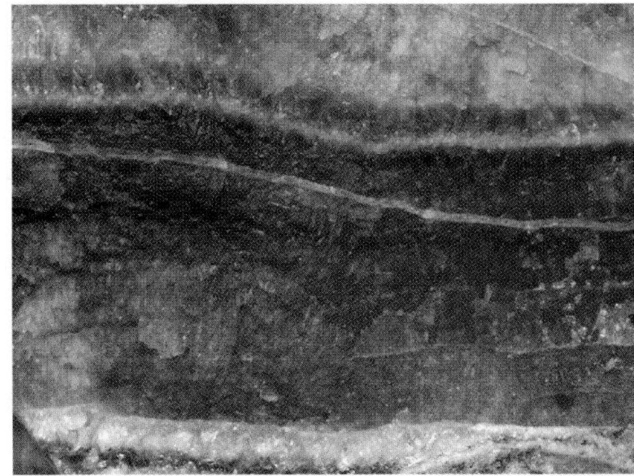

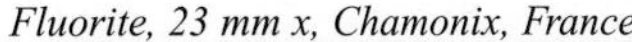

Fluorite, 23 mm x, Chamonix, France

4. Oxides

Cuprite

Cu_2O

CUBIC ● ● ●

Varieties: chalcotrichite (acicular to hair-like crystals)
Properties: C – red; S – red; L – adamantine to sub-metallic; D – transparent to translucent; DE – 5.8-6.2; H – 3.5-4; CL – imperfect; F – conchoidal to uneven; M – combinations of cubic crystals, hair-like aggregates, granular, massive.
Origin and occurrence: Secondary, as a result of the oxidation of Cu sulfides. Crystals up to 150 mm (6 in) in size, covered with malachite, occur in Onganja, Namibia. Shiny octahedra, up to 40 mm ($1^{9}/_{16}$ in) in size, come from the Mashamba West mine, Zaire. Acicular and fibrous crystals of the chalcotrichite variety were found in Bisbee, Arizona, USA. Combinations of cubic crystals, covered with malachite, are known from Chessy near Lyon, France. Crystals up to 30 mm ($1^{3}/_{16}$ in) across are reported from Tsumeb, Namibia.
Application: important Cu ore.

Zincite

ZnO

HEXAGONAL ● ●

Properties: C – yellow, orange, red; S – orange-yellow; L – adamantine; D – transparent to translucent; DE – 5.7; H – 4.5-5; CL – perfect; F – conchoidal; M – pyramidal crystals, granular, massive.
Origin and occurrence: Metamorphic, associated with willemite and franklinite. It forms very rare crystals up to 40 mm ($1^{9}/_{16}$ in) in size in the metamorphosed Zn deposits in Franklin and Sterling Hill, New Jersey, USA, it is mostly granular and massive. Zincite crystals and aggregates of vitreous luster from Poland, which are offered at the mineral shows, are not of a natural origin, there are smelter products.

Zincite, 11 mm x, Franklin, U.S.A.

Amethyst, 122 mm, Guerrero, Mexico
Cuprite, 18 mm x, Mashamba West, Zair

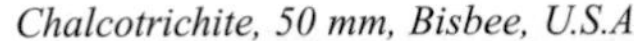

Chalcotrichite, 50 mm, Bisbee, U.S.A.

Tenorite
CuO

MONOCLINIC ● ● ●

Properties: C – steel-gray to black; S – gray; L – metallic; D – opaque; DE – 6.5; H – 3.5; CL – imperfect; F – uneven to conchoidal; M – thin tabular to scaly crystals, earthy, massive.
Origin and occurrence: Secondary, in the oxidation zone of Cu deposits, together with other Cu supergene minerals. It was common in Cu deposits in the Keweenaw Peninsula, Michigan, USA. It was mined as Cu ore in Bisbee, Globe and Morenci, Arizona, USA. Thin tabular crystals are known from Tsumeb, Namibia.

Spinel
$MgAl_2O_4$

CUBIC ● ● ●

Varieties: pleonast (black)
Properties: C – pink, red, green, blue, brown, black; S – white; L – vitreous to dull; D – transparent to opaque; DE – 3.6; H – 7.5-8; CL – imperfect; F – conchoidal to uneven; M – octahedral crystals, granular, massive.
Origin and occurrence: Magmatic, metamorphic, also in placers, associated with corundum, sillimanite and other minerals. Large pleonast crystals reaching up to 150 mm (6 in) were found in the Aldan massif, Yakutia,

Tenorite, 3 mm xx, Bisbee, U.S.A

Russia. Fine crystals weighing up to 14 kg (30 lb 12 oz) come from Amity, New York. Crystals up to 120 mm (4 $^{7}/_{16}$ in) across are known from Sterling Hill, New Jersey. Blue crystals of spinel come from Bolton, Massachusetts, USA and South Burgess, Ontario, Canada. Gemmy pink and red crystals up to 20 mm ($^{25}/_{32}$ in) reported near Ratnapura, Sri Lanka and in Mogok, Burma. Fine pink crystals up to 50 mm (2 in) in size occur in Kukh-i-lal, Tajikistan.
Application: gemstone.

Gahnite
$ZnAl_2O_4$

CUBIC ● ● ●

Properties: C – black-green, black; S – gray; L – vitreous to greasy; D – translucent to opaque; DE –

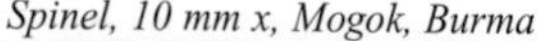

Spinel, 10 mm x, Mogok, Burma

Pleonast, 15 mm xx, Vietnam

Gahnite, 90 mm, Broken Hill, Australia

4.4-4.6; H – 7.5-8; CL – imperfect; F – conchoidal to uneven; M – octahedral crystals, granular.

Origin and occurrence: Magmatic and metamorphic, associated with wolframite, chalcopyrite and other minerals. Crystals up to 120 mm ($4^{7}/_{16}$ in) in size come from Franklin and Sterling Hill, New Jersey, USA. Crystals from Broken Hill, New South Wales, Australia reached up to 30 mm ($1^{3}/_{16}$ in). Blue-green crystals were found in Rowe, Massachusetts, USA. Cuttable blue crystals occur near Gidan Wayo, Nigeria.

Magnetite

$Fe^{2+}Fe^{3+}{}_2O_4$

CUBIC ● ● ●

Properties: C – black; S – black; L – metallic; D – opaque; DE – 5.2; H – 5.5-6.5; CL – none; F – uneven to conchoidal; M – octahedral crystals, granular, massive.

Origin and occurrence: Magmatic, hydrothermal and metamorphic, rare sedimentary. Parageneses differ according to the origin. Fine crystals up to 170 mm ($6^{11}/_{16}$ in) in size, come from Traversella, Italy. A crystal 25 cm ($9^{13}/_{16}$ in) in size was found in Vastanfors, Sweden. Fine crystals up to 40 mm ($1^{9}/_{16}$ in) occur in Dashkesan, Azerbaijan, where it is associated with andradite, epidote and apatite. Beautiful shiny octahedra up to 40 mm ($1^{9}/_{16}$ in) are known from Alpa Lercheltini, Binntal, Switzerland. Rare cubes up to 20 mm ($^{25}/_{32}$ in) on edge come from the ZCA No.4 mine, Balmat, New York, USA. Magnetite crystals reaching up to 10 cm were found in pegmatites in Jaguaraçu, Minas Gerais, Brazil. Crystals up to 20 cm ($7^{7}/_{8}$ in) in size reported from the Gardiner complex, Greenland. Crystals up to 50 mm (2 in) were lately found in Kovdor, Kola Peninsula, Russia.

Application: Fe ore.

Magnetite, 20 mm x, Chester, U.S.A.

Franklinite, 30 mm, Sterling Hill, U.S.A.

Franklinite

$(Zn,Mn^{2+},Fe^{2+})(Fe^{3+},Mn^{2+})_2O_4$

CUBIC ● ●

Properties: C – black; S – dark brown; L – metallic; D – opaque; DE – 5.1-5.5; H – 6; CL – none; F – uneven to conchoidal; M – octahedral crystals, granular, massive.
Origin and occurrence: Metamorphic, associated with willemite, zincite and other minerals. The only localities, where it is common and occurs in very large accumulations, are Franklin and Sterling Hill, New Jersey, USA. Crystals up to 170 mm (6 11/16 in) across are known from there. It is rare in Langban, Sweden and Ocna de Fier, Romania.
Application: Zn ore.

Chromite

$FeCr_2O_4$

CUBIC ● ● ●

Properties: C – black; S – brown; L – metallic; D – opaque; DE – 4.5-4.8; H – 5; CL – none; F – uneven; M – octahedral crystals, granular, massive.
Origin and occurrence: Magmatic, together with magnetite, uvarovite and other minerals. Rare crystals reaching up to 10 mm (3/8 in) are known from Uzun Damar, Turkey. It occurs mostly massive, like in deposits in Bushveld, South Africa; in Sarany, Ural Mountains, Russia; and in Guleman, Turkey.
Application: Cr ore.

Hausmannite

$Mn^{2+}Mn^{3+}{}_2O_4$

TETRAGONAL ● ● ●

Properties: C – black; S – brown; L – submetallic; D – opaque; DE – 4.8; H – 5.5; CL – perfect; F – uneven; M – pseudo-octahedral crystals, granular, massive.
Origin and occurrence: Hydrothermal in high-temperature Mn deposits, also as a product of the contact metamorphism. The best specimens with crystals up to 30 mm (1 3/16 in) in size come from the N'Chwaning mine, Kuruman, South Africa. Smaller crystals were found in Ilfeld and Ilmenau, Germany. It also occurred as fine crystals in Langban and Jakobsberg, Sweden.

Chromite, 87 mm, Finnland

Minium, 60 mm, Broken Hill, Australia

Minium
$Pb^{2+}{}_2Pb^{4+}O_4$

TETRAGONAL ● ● ●

Properties: C – red, S – orange-yellow; L – dull to greasy; D – opaque; DE – 8.9; H – 2.5; CL – perfect; M – earthy and pulverulent aggregates, massive.

Origin and occurrence: Secondary mineral, as a result of the galena oxidation. It occurs in Langban, Sweden; in Anarak, Iran; in Leadhills, Scotland; and in Broken Hill, New South Wales, Australia.

Hausmannite, 42 mm, Kuruman, South Africa

Chrysoberyl, 35 mm, Espírito Santo, Brazil

Chrysoberyl

$BeAl_2O_4$

ORTHORHOMBIC ● ● ●

Varieties: alexandrite

Properties: C – yellow-green, yellow, blue-green, alexandrite is green in daylight, purple in artificial light; S – white; L – vitreous; D – transparent to translucent; DE – 3.8; H – 8.5; CL – good; F – conchoidal to uneven; M – thin to thick tabular crystals common cyclic twins.

Origin and occurrence: Magmatic in pegmatites, prevailing as metamorphic, in association with schorl, phenakite and other minerals. Twins up to 22 cm ($8^{11}/_{16}$ in) occurred near Pancas, Espírito Santo, Brazil. Complicated twins up to 100 mm (4 in) in

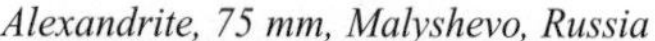

Alexandrite, 75 mm, Malyshevo, Russia

size come from Ambatondrazaka and other localities in Madagascar. Tabular crystals several cm in size, embedded in the sillimanitic rock, were found in Marsíkov, Czech Republic. Fine alexandrite crystals up to 80 mm (3 1/8 in) are known from Malyshevo, Ural mountains, Russia, together with emerald and phenakite. Alexandrite crystals reach up to 30 mm (1 3/16 in) in Nyanda, Zimbabwe. Gemmy chrysoberyls, commonly with a cat's eye effect, come from the vicinity of Ratnapura, Sri Lanka.
Application: gemstone.

Valentinite
Sb_2O_3

ORTHORHOMBIC ● ● ●

Properties: C – colorless, white, brownish; S – white; L – adamantine; D – transparent to translucent; DE – 5.7-5.8; H – 2.5-3; CL – perfect; M – prismatic to tabular crystals, radial aggregates, massive.
Origin and occurrence: Secondary mineral, originated in the oxidation of stibnite. The best specimens with crystals up to 30 mm (1 3/16 in) were found in Příbram, Czech Republic. Fine crystals come also from Bräunsdorf, Germany. Crystals up to 20 mm (25/32 in) long occur in Oruro, Bolivia. Beautiful radial aggregates up to 40 mm (1 9/16 in) in diameter, associated with kermesite, are known from Pezinok and Pernek, Slovakia. Pseudo-morphs after stibnite crystals up to 35 cm (13 6/8 in) long, are reported from the Xikuangshan Mine, Lengshuijiang, China.

Valentinite, 40 mm, Nicolet, Canada

Arsenolite
As_2O_3

CUBIC ● ●

Properties: : C – white; S – white; L – vitreous; D – transparent to translucent; DE – 3.9; H – 1.5; CL – good; F – conchoidal; M – octahedral crystals, crusts, coatings; R – soluble in water.
Origin and occurrence: Secondary mineral, resulting from the oxidation of As ores. Poorly developed crystals several mm long, occur in Jáchymov, Czech Republic, in Johanngeorgenstadt and St. Andreasberg, Germany. Crystals up to 20 mm (25/32 in) long originated during a mine fire in the White Caps mine, Nevada, USA.

Arsenolite, 3 mm xx, Récsk, Hungary

Senarmontite
Sb_2O_3

CUBIC ● ● ●

Properties: C – white, light gray; S – white; L – greasy, vitreous to adamantine; D – transparent to opaque; DE – 5.2-5.8; H – 2-2.5; CL – imperfect; F – uneven; M – octahedral crystals, granular, massive.
Origin and occurrence: Secondary, produced by stibnite oxidation, with valentinite and cerussite. The finest crystals up to 30 mm (1 3/16 in) in size come from Djebel Hammimate, Algeria. Also occurred in Cetine, Italy and in Dúbrava, Slovakia.

Senarmontite, 3 mm xx, Pernek, Slovakia

Bixbyite, 11 mm x, Thomas Range, U.S.A.

Leucosapphire, 30 mm, Sri Lanka

Bixbyite
$(Mn^{3+},Fe^{3+})_2O_3$

CUBIC ● ●

Properties: C – black; S – black; L – metallic to submetallic; D – opaque; DE – 5; H – 6-6.5; CLB imperfect; F – conchoidal to uneven; M – cubic crystals, also twins.
Origin and occurrence: Hydrothermal in rhyolite cavities and metamorphic. Cubes up to 12 mm (15/32 in) are found together with topaz in Thomas Range, Utah, USA. Crystals up to 25 mm (1 in) occurred in the Postmasburg mine, South Africa. Crystals up to 80 mm (3 1/8 in) come from Ultevis, Sweden.

Corundum
Al_2O_3

TRIGONAL ● ● ● ●

Varieties: ruby, sapphire, leucosapphire, emery

Properties: C – colorless (leucosapphire), yellow, pink, red (ruby), blue (sapphire), purple, green,

Corundum, 10 mm xx, Miass, Russia

Ruby, 15 mm xx, Kola, Russia

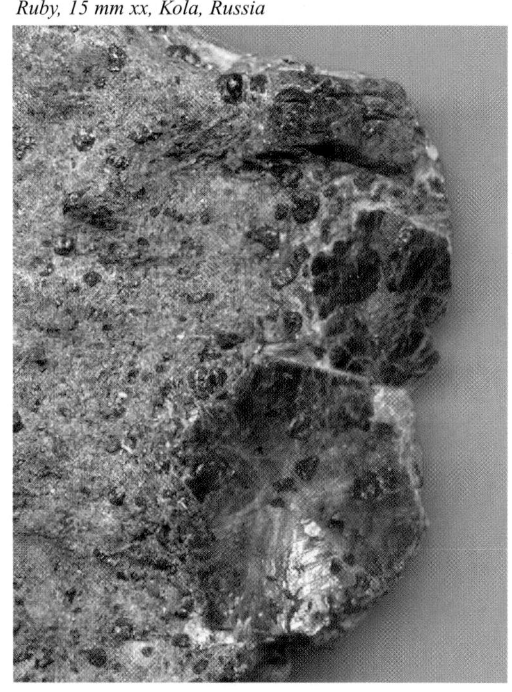

Corundum, 4 mm, Montana, U.S.A.

Ruby, 16 mm x, Jegdalek, Afghanistan

gray; S – white; L – vitreous to adamantine; D – transparent to opaque; DE – 4.0-4.1; H – 9; CL – none; F – conchoidal to uneven; M – long prismatic to barrel-like crystals, pebbles; LU – rare dark red.

Origin and occurrence: Magmatic in andesites, pegmatites and basalts, metamorphic and in placers, in association with andalusite, topaz, spinel and other minerals. Crystals of common corundum weighed up to 30 kg (66 lb) near Bancroft, Ontario, Canada. A crystal, weighing 151 kg (333 lb 3 oz) was also found in the Letaba district, South Africa. Sapphire crystals weighing up to 20 kg (44 lb), come from the vicinity of Ratnapura and Rakwana, Sri Lanka. Fine sapphire crystals are also known from Kashmir, India. Rough gem sapphire is mined from the Yogo Gulch sediments in Montana, USA and from Anikia, Queensland, Australia. Fine blue crystals up to 50 mm (2 in) long, occur near Miass, Ural mountains, Russia. Ruby is even much rarer variety of corundum. Its beautiful crystals up to 50 mm (2 in) long come from Jegdalek, Afghanistan; Mogok, Burma and from Luc Yen, Vietnam. Prismatic crystals of opaque ruby, up to 40 mm (1 9/16 in) in size were found in the Khit Island near Kola Peninsula, Russia. Ruby crystals up to 30 cm (12 in) in size embedded in green zoisite from the vicinity of Arusha, Tanzania are very decorative.

Application: emery as abrasive material, sapphire and ruby as gemstones.

Sapphire, 41 mm, Kashmir, India

Hematite, 50 mm, La Fibbia, Switzerland

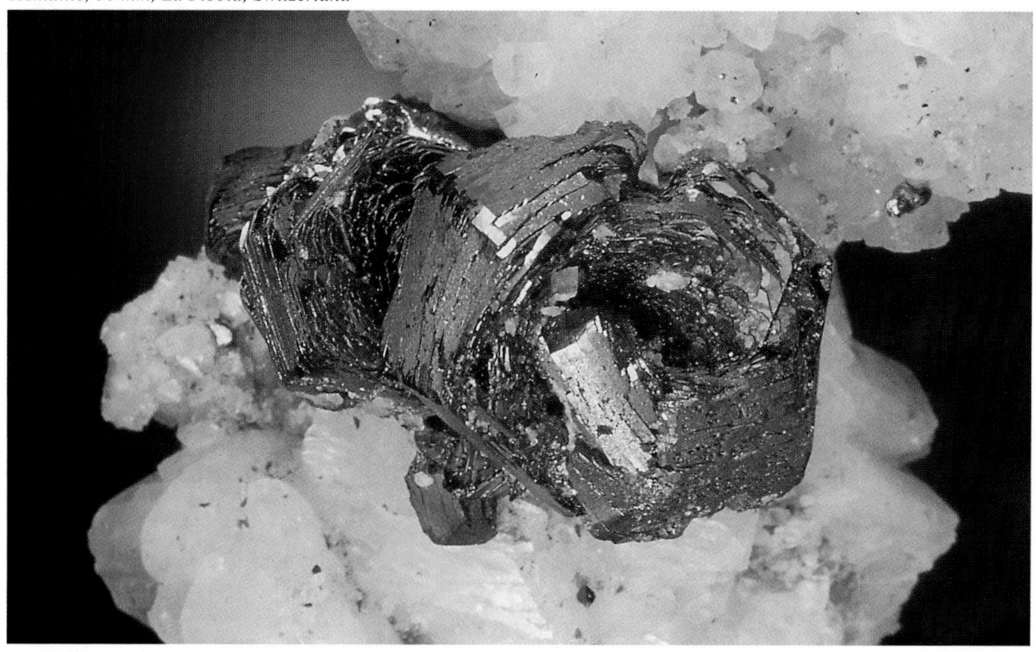

Hematite

Fe_2O_3

TRIGONAL ● ● ● ● ●

Properties: C – red, gray, black; S – red; L – metallic, dull; D – opaque; DE – 5.3; H – 6-6.5. earthy to 1; CL – none; F – uneven to conchoidal; M – thick to thin tabular crystals, massive, earthy.
Origin and occurrence: Magmatic, hydrothermal, sedimentary, also metamorphic, parageneses vary according to the origin. Beautiful crystals up to 100 mm (4 in) in size come from Brumado, BahRa, Brazil. Crystals up to 30 cm (12 in), were found in the Wessels mine, Kuruman, South Africa. So called iron roses reached up to 100 mm (4 in) near St. Gotthard, Switzerland. Fine crystals several cm in size, occurred in Rio Marina, Elba, Italy. Very fine tabular crystals reached up to 70 mm (2¾ in) in Nador, Morocco. New finds of fine crystals up to 40 mm (1⁹⁄₁₆ in) in size were made in the Korshunovskoye deposit, Russia. Fine botryoidal aggregates come from Hradište and Horní Blatná, Czech Republic and from Botallack, Cornwall, UK. Sedimentary banded iron ores form huge deposits near Krivoy Rog, Ukraine or in the vicinity of Lake Superior (Mesabi Range, Minnesota; Marquette, Michigan, USA).
Application: important Fe ore.

Loparite-(Ce), 10 mm xx, Khibiny Massif, Kola, Russia

Ilmenite

$FeTiO_3$

TRIGONAL ● ● ● ●

Properties: C – black; S – black; L – metallic to dull; D – opaque; DE – 4.5-5; H – 5-6; CL – none; F – conchoidal to uneven; M – thick tabular crystals, granular, massive.
Origin and occurrence: Magmatic, metamorphic and in placers, associated with pyrrhotite, rutile, magnetite and other minerals. Crystals weighing up

Ilmenite, 11 mm x, Binntal, Switzerland

to 30 kg (66 lb) were described from the Faraday mine near Bancroft, Ontario, Canada. Crystals up to 150 mm (6 in) in size occurred near Girardville, Quebec, Canada. Crystals also reached up to 100 mm (4 in) near Miass, Ural mountains, Russia. Crystals up to 120 mm 120 mm (4 7/16 in), were found in Arendal and Kragerö, Norway. Crystal rosettes up to 10 mm (3/8 in) in size come from Maderanertal, Switzerland. It is also common in placers (Kamituga, Kivu, Zaire; Sri Lanka; Travancore, India; Madagascar etc.).

Perovskite

$CaTiO_3$

ORTHORHOMBIC ● ● ●

Properties: C – dark brown to black; S – colorless to gray; L – metallic to adamantine; D – opaque; DE – 4.0-4.3; H – 5.5-6; CL – imperfect; F – conchoidal to uneven; M – pseudo-cubic crystals, granular.
Origin and occurrence: Magmatic in basic and ultrabasic rocks, metamorphic, together with magnetite, zircon and other minerals. Fine pseudo-cubic crystals up to 40 mm (1 9/16 in) in size come from Zlatoust and Akhmatovsk, Ural mountains, Russia. It occurs as crystals up to 80 mm (3 1/8 in), associated with magnetite crystals in the Gardiner complex, Greenland. Crystals up to 40 mm (1 9/16 in) were found in Jacupiranga, Sao Paulo, Brazil. Crystals from Val Malenco, Italy, reached up to 20 mm (25/32 in). Crystals up to 20 mm (25/32 in) were lately found in Afrikanda, Kola Peninsula, Russia.

Loparite-(Ce)

$(Ce,Na,Ca)TiO_3$

ORTHORHOMBIC ●

Properties: C – black, S – dark red-brown, L – metallic; D – opaque; DE – 4.6-4.9; H – 5.5-6; M – pseudo-cubic crystals, granular; R – metamict.
Origin and occurrence: Magmatic in alkaline rocks, with lorenzenite, eudialyte and aegirine. Fine interpenetration twins up to 20 mm (25/32 in) in size come from Mount Nyorkpakhk, Kola Peninsula, Russia.

Perovskite, 60 mm, Zlatoust, Russia

Stibiconite, 230 mm, Catorce, Mexico

Stibiconite
$Sb^{3+}Sb^{5+}{}_2O_6(OH)$

CUBIC ● ● ●

Properties: C – white, creamy, light yellow, brown; S – white; L – vitreous, greasy to dull; D – opaque; DE – 4.1-5.8; H – 3-6; M – pseudo-morphs after stibnite crystals, earthy, massive.
Origin and occurrence: Secondary, as a result of the stibnite oxidation, associated with valentinite and other minerals. Fine pseudo-morphs after stibnite crystals up to 30 cm (12 in) long come from Catorce, San Luis Potosí, Mexico. Similar pseudo-morphs were also found in Kostainik, Serbia; in the Ichinokawa mine, Japan and in Pereta, Italy. Pseudo-morphs after stibnite up to 20 cm (7 7/8 in) long occur also in Çukurören, Turkey.

Bindheimite
$Pb_2Sb_2O_6(O,OH)$

CUBIC ● ● ●

Properties: C – yellow, brown, gray; S – yellow, L – resinous, dull to earthy; D – translucent to opaque; DE – 4.6-5.6; H – 4-4.5; F – conchoidal to earthy; M – botryoidal, nodular and earthy crusts.
Origin and occurrence: Secondary in the oxidation zone of Pb-Sb deposits. Needles up to 10 mm (3/8 in) long come from Rudník, Czech Republic. It is common in Broken Hill, New South Wales, Australia; in Bisbee, Arizona, USA and in Sidi-Amor-ben-Salem, Tunisia. Lamellar pseudo-morphs up to several cm in size are known from Tsumeb, Namibia.

Pyrochlore
$(Na,Ca)_2Nb_2O_6(OH,F)$

CUBIC ● ● ●

Properties: C – yellow-brown, brown, black; S – brown; L – vitreous to greasy; D – translucent to opaque, DE – 4.5; H – 5-5.5; CL – locally good; F – conchoidal to uneven; M – octahedral crystals, granular; R – radioactive (admixtures of U, Th).
Origin and occurrence: Magmatic in alkaline rocks, together with zircon, astrophyllite and other minerals. Fine brown shiny crystals up to 20 mm (25/32 in) in size come from the vicinity of Vishnevogorsk, Ural mountains, Russia. Crystals reaching 10 mm (3/8 in) occur in the Panda Hill deposit, Tanzania. Crystals are also known from Oka, Quebec, Canada. Single octahedra measuring 5 mm (3/16 in) were found in Luesha, Kivu, Zaire. *Application:* Nb,U and Th ore.

Bindheimite, 4 mm, Rudabánya, Hungary

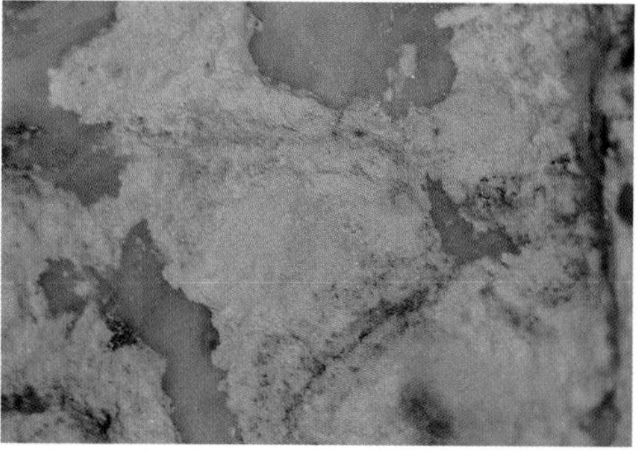

Microlite, 7 mm x, Gillette Quarry, U.S.A.

Pyrochlore, 58 mm, Vishnevogorsk, Ural Mts., Russia

Betafite

$(Ca,Na,U)_2(Ti,Nb,Ta)_2O_6(OH)$

CUBIC ● ●

Properties: C – black, brown, yellow-brown; SB red-brown; L – resinous to greasy; D – translucent to opaque; DE – 4.2; H – 3-5.5; CL – none; F – conchoidal to uneven; M – octahedral crystals; R – radioactive, metamict.
Origin and occurrence: Magmatic in granitic pegmatites, rich in U, Th and rare earth elements, associated with beryl, euxenite-(Y) and other minerals.

Betafite, 20 mm, Silver Crater, Canada

The world's best specimens come from many localities in Madagascar (Betafo, Ambatofotsikely etc.), where crystals up to 6 kg (13 lb 3 oz) were found. Beautiful specimens with crystals up to 100 mm (4 in) in size occur in the Silver Crater mine near Bancroft, Ontario, Canada. It is also known from Evje, Norway.

Microlite

$(Na,Ca)_2Ta_2O_6(O,OH,F)$

CUBIC ● ● ●

Properties: C – brown, yellow, green, reddish; S – white; L – vitreous to greasy, locally adamantine; D – translucent to opaque; DE – 5-6.4; H – 6-6.5; CL – locally good; F – conchoidal to uneven; M – octahedral crystals, granular, massive.
Origin and occurrence: Magmatic, typical for granitic pegmatites, together with manganocolumbite, manganotantalite and other minerals. Octahedra up to 65 mm ($2\frac{9}{16}$ in) in size occur in Ankola, Uganda. Crystals up to 30 mm ($1\frac{3}{16}$ in) in size come from Virgem da Lapa, Minas Gerais, Brazil. Crystals up to 75 mm (3 in) are reported from the Harding pegmatite, New Mexico, USA. It occurs in important accumulations near Wodgina, Western Australia.

Quartz, 60 mm, Herkimer, U.S.A.

Quartz

SiO_2

TRIGONAL ●●●●●

Varieties: rock crystal, citrine, smoky citrine, morion, amethyst, rose quartz, chrysoprase, jasper, chalcedony, agate, onyx, sardonyx, aventurine, heliotrope, tiger's eye, falcon's eye.

Properties: C – colorless (rock crystal), white, yellow (citrine), brown (smoky citrine), black (morion), purple (amethyst), pink (rose quartz), green (chrysoprase); D – these varieties are mostly transparent, often translucent; C – other varieties are mainly multicolored, separate colors have different hues and the color is commonly caused by microscopic admixtures of other minerals; varieties: red, green, brown, yellow (jasper), banded with different colors (agate); white

Rock crystal, 95 mm, La Gardette, France

Citrine, 32 mm, Charcas, Mexico

and black bands (onyx), white and red-brown bands (sardonyx), green to red-brown with mica or hematite inclusions (aventurine), dark green with red spots (heliotrope), yellow-brown to black-brown, fibrous with silky luster (tiger's eye), blue-gray to yellow-brown, fibrous with silky luster (falcon's eye); S – white; L – vitreous, silky, dull; D – transparent to translucent, opaque; DE – 2.6; H – 7; CL – none; F –

Smoky quartz, 70 mm, Middle Moat Mt., U.S.A.

Smoky quartz, 81 mm, Switzerland

Morion, 100 mm, Agadir, Kazakhstan

Amethyst, 100 mm, Bochovice, Czech Republic

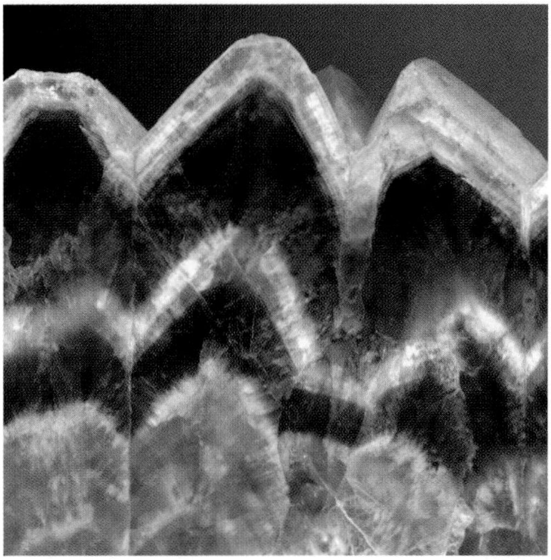

conchoidal; M – long to short prismatic, acicular, dipyramidal to tabular crystals, fibrous, botryoidal and stalactitic aggregates and coatings, concretions, geodes, granular, massive.

Origin and occurrence: Magmatic in different types of rocks, mainly in granites, granitic pegmatites and volcanic rocks; metamorphic in different types of rocks, mainly in quartzites and mica schists; hydrothermal in different types of ore and Alpine-

Rose quartz, 145 mm, Minas Gerais, Brazil

Amethyst, 200 mm, Guerrero, Mexico

type veins; secondary in the oxidation zone of ore deposits; also in different types of sedimentary rocks and in organic remains, also in placers. Probably the most common mineral in the Earth's crust and the most important rock-forming mineral, as well. Large crystals of rock quartz up to 7 m (23 ft) long come from pegmatites in the Betafo region in Madagascar and from the Alpine-type veins, like Uri, Grimsel and Furka, Switzerland; perfect crystals are known from the cracks in marbles near Carrara, Italy; it also occurs in the quartz veins in Herkimer, New York and Hot Springs, Arkansas, USA. Citrine occurs mainly in granitic pegmatites and large crystals come from Goiás, Brazil; from Suky and Netín, Czech Republic; from Murzinka, Ural mountains, Russia. Smoky quartz originates mostly in granitic pegmatites, it also occurs in the Alpine-type quartz veins and in cavi-

Chalcedony, 68 mm, High Atlas, Morocco

Carnelian, 40 mm, Nová Paka, Czech Republic

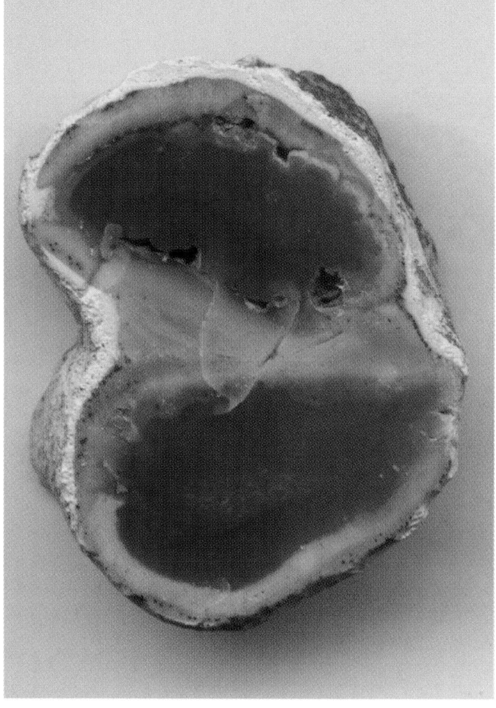

Chrysoprase, 60 mm, Szklary, Poland

ties of volcanic rocks. Perfect crystals up to several meters long, come from many places, the largest crystal, weighing 77 tons, was found in Kazakhstan; perfect crystals occur in pegmatites in many places in Brazil; also in Korostenskiy massif, Ukraine; in the Pikes Peak batholith, Colorado, USA; in the Alpine-type veins in Maderanertal; and in Grimsel, Switzerland. Morion crystals, commonly associated with smoky citrine, were found in quartz veins and in pegmatites. Its crystals are known from St. Gotthard, Switzerland. Amethyst comes from quartz and ore veins, cavities in volcanic rocks, rare in the Alpine-type veins. Famous localities in volcanic rocks are in the states of Rio Grande do Sul and Minas Gerais, Brazil, doubly terminated crystal, weighing 5.5 tons, come from Diamantina. In Serra do Mar, Rio Grande do Sul, a cavity covered with amethyst crystals measuring 10 x 2 x 1 m (33 x 6 x 3 ft 3in) was found; rich druses occur also

Jasper, 50 mm, Ural Mts., Russia

Jasper, 60 mm, Oregon, U.S.A.

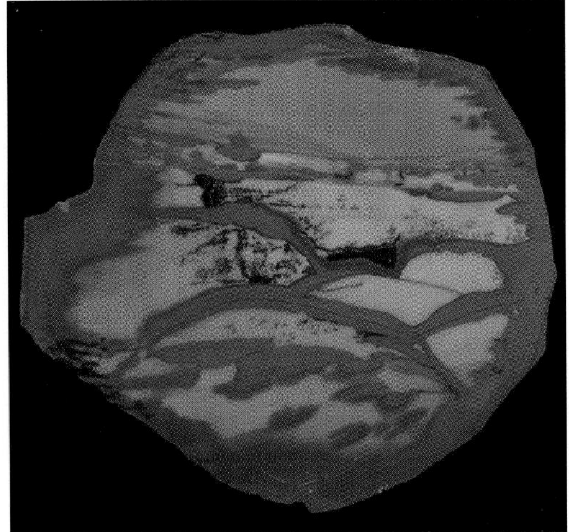

Petrified wood, 70 mm, Podkrkonoší, Czech Republic

in the ore veins in Porcura, Romania and Julimes, Mexico. Rose quartz, forming masses up to several meters in granitic pegmatites in the Rose Quartz pit, Quadeville, Ontario, Canada; in Ambositra, Madagascar; crystals up to 10 mm (3/8 in) long growing on quartz crystals, come from Sapucaia, Minas Gerais, Brazil. Dark green chrysoprase veins up to 50 mm (2 in) thick are known from serpentinites in Szklary, Poland. Jasper is known from volcanic rocks and their contacts with sediments,

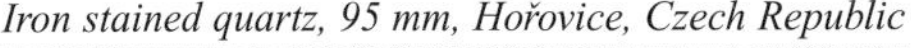

Iron stained quartz, 95 mm, Hořovice, Czech Republic

Agate, 50 mm, Železnice, Czech Republic

Onyx, 65 mm, Brazil

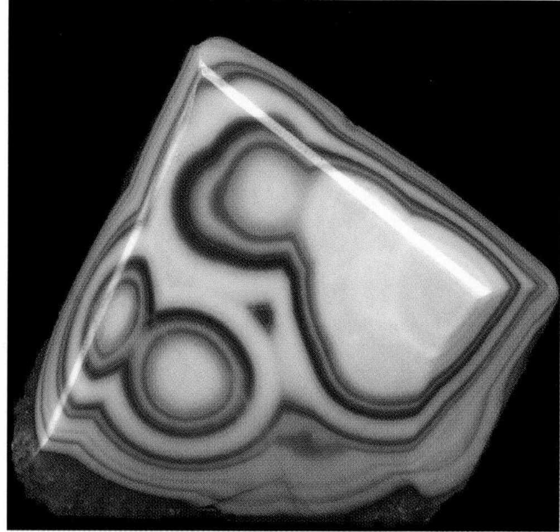

locally as a result of petrification of organic matter, mainly plants, it is also known from quartz veins. Rich aggregates occur in Idar-Oberstein, Germany; in Podkrkonoší region and in Krušné hory mountains, Czech Republic; in the Petrified Forest National Park, Holbrook, Arizona, USA; in Ural mountains, Russia; and in Kabamby, Madagascar. Chalcedony is mostly found in quartz veins and geodes in volcanic rocks, also in sediments. Rich aggregates come from Idar-Oberstein, Germany; Julimes, Mexico; many localities in Uruguay and Brazil; and in Hüttenberg, Austria. Agates are known from cavities in volcanic rocks, rare in hydrothermal veins and in sediments. The most important localities are located in the southern part of Brazil in the state of Rio Grande do Sul; in Uruguay; also in Yemen; India; Mongolia; in several localities in the USA; in Idar-Oberstein, Germany; and in Podkrkonoší, Czech Republic. The most

Moss agate, 40 mm, Krdjali, Bulgaria

Agate, 140 mm, Horní Halže, Czech Republic

Agate, 60 mm, Brazil

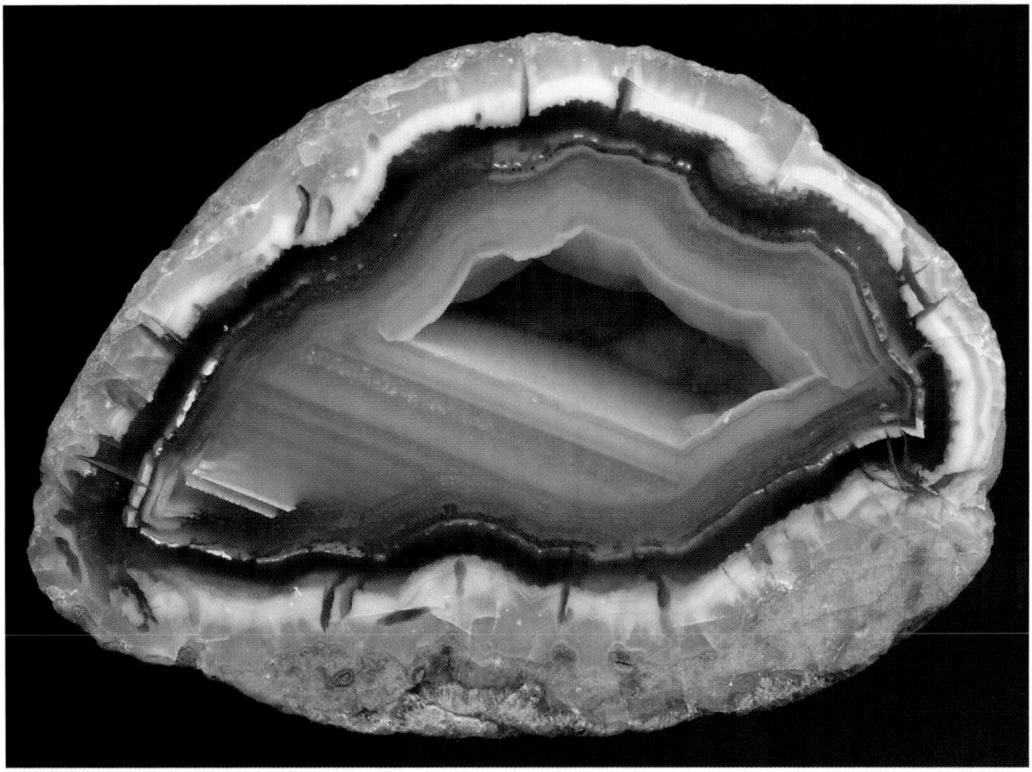

famous localities of onyx and sardonyx are in Brazil and Uruguay. Rich aggregates of aventurine come from Miass, Ural mountains, Russia; Mariazell, Austria; Belany, India. Heliotrope occurs in Idar-Oberstein, Germany; Kozákov, Czech Republic; and in Brazil.
Application: important raw material in glass industry, many colored varieties, like amethyst, smoky citrine, citrine, onyx, sardonyx, and heliotrope are cut as gemstones.

Aventurine, 50 mm, India

Tiger´s eye, 50 mm, Griqualand, South Africa

Tridymite, 9 mm, Big Luc Mt., U.S.A.

Tridymite
SiO_2

ORTHORHOMBIC ● ● ●

Properties: C – colorless to white; S – white; L – vitreous; D – transparent; DE – 2.3; H – 6.5-7; CL – none; F – conchoidal; M – pseudo-hexagonal tabular crystals.

Origin and occurrence: Magmatic in cavities of young felsic volcanic rocks in association with cristobalite, chalcedony and other minerals. Pseudo-hexagonal tabular crystals up to 10 mm ($^{3}/_{8}$ in) in size, come from Vechec, Slovakia. Similar crystals found in Ichigayama, Japan. Crystals up to 10 mm ($^{3}/_{8}$ in) occur with topaz and other minerals in the Thomas Range, Utah, USA.

Cristobalite, 110 mm, Vechec, Slovakia

Cristobalite
SiO_2

TETRAGONAL ● ● ●

Varieties: lussatite (fibrous)

Properties: C – colorless to white; S – white; L – vitreous; D – translucent; DE – 2.3; H – 6.5; CL – none; M – pseudo-octahedral crystals, spherical and botryoidal aggregates.
Origin and occurrence: Magmatic in cavities of young felsic volcanic rocks, associated with tridymite. Crystals up to 4 mm ($^{5}/_{32}$ in) are known from Cerro San Cristobal, Hidalgo, Mexico. Crystals up to 2 mm ($^{1}/_{16}$ in) long occur in Vechec, Slovakia. Gray spherical aggregates come from Coso Hot Springs, California, USA.

Common opal, 60 mm, Křemže, Czech Republic

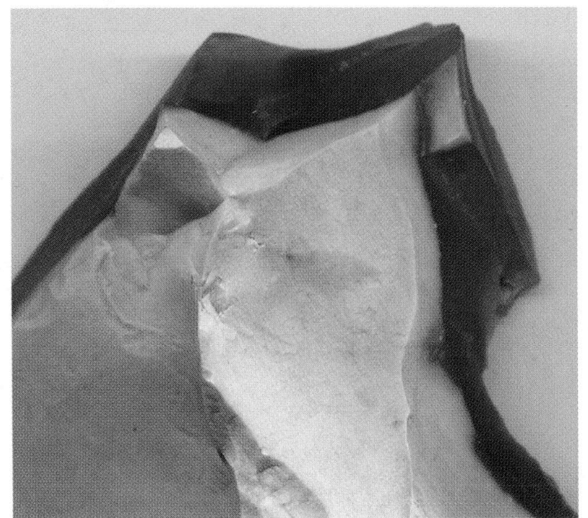

Opal, 50 mm, Herlany, Slovakia

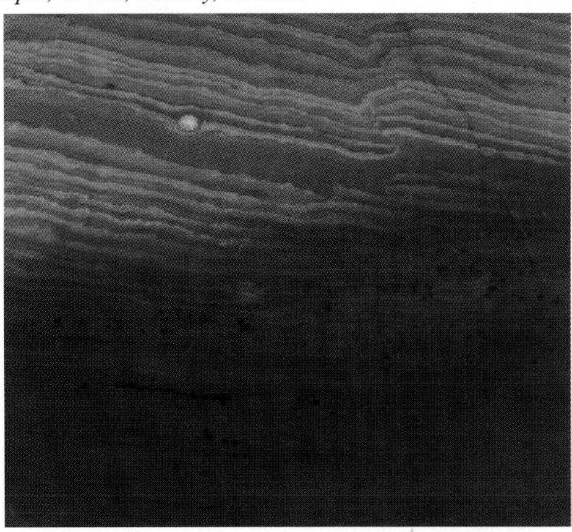

Opal

$SiO_2 . n H_2O$

AMORPHOUS ● ● ● ● ●

Varieties: hyalite, milky opal, fire opal, precious opal, wooden opal, geysirite.

Properties: C – colorless (hyalite), white (milky opal), red (fire opal), iridescence (precious opal), brown, red-brown, yellow, green, gray, blue; S – white; L – vitreous, dull, earthy, waxy; D – transparent to translucent, opaque; DE – 2.1; H – 5.5-6.5; CL – none; F – conchoidal; M – botryoidal and stalactitic aggregates, coatings, concretions, geodes, massive; LU – white, yellow-green, green.

Origin and occurrence: Hydrothermal in volcanic rocks and tuffs, also in various types of volcanic rocks and tuffites, in different types of sedimentary rocks, in organic remnants and hot springs, rare in

Hyalite, 45 mm, Valeč, Czech Republic

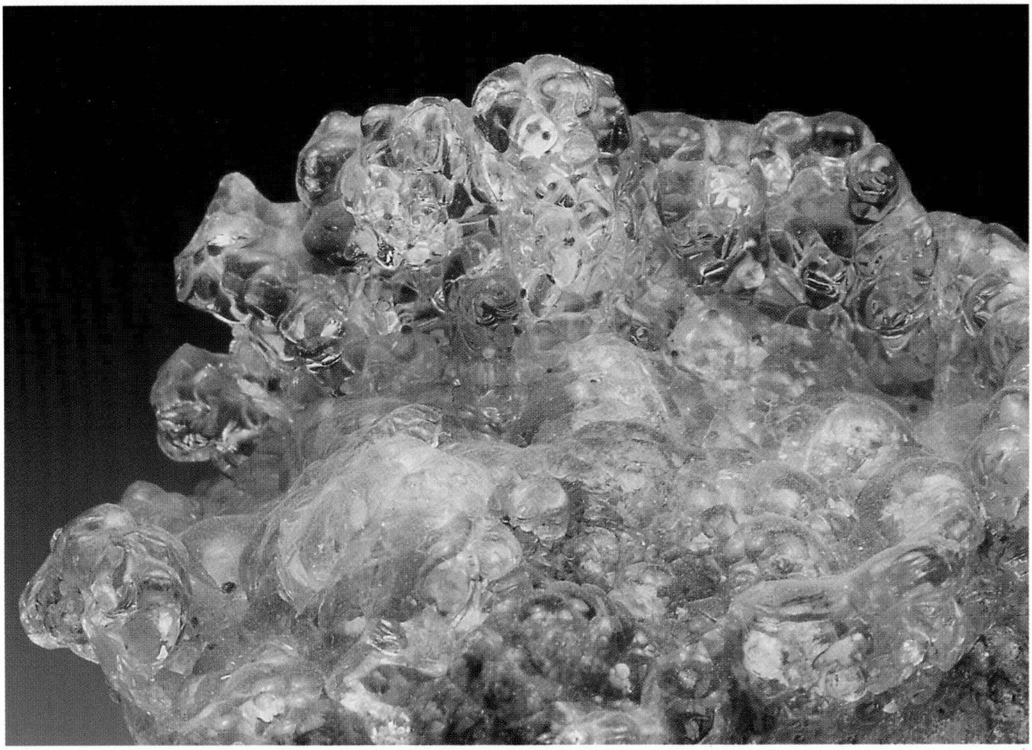

Opal, 47 mm, Opal Butte, U.S.A.

Wooden opal, 40 mm, Lubietová, Slovakia

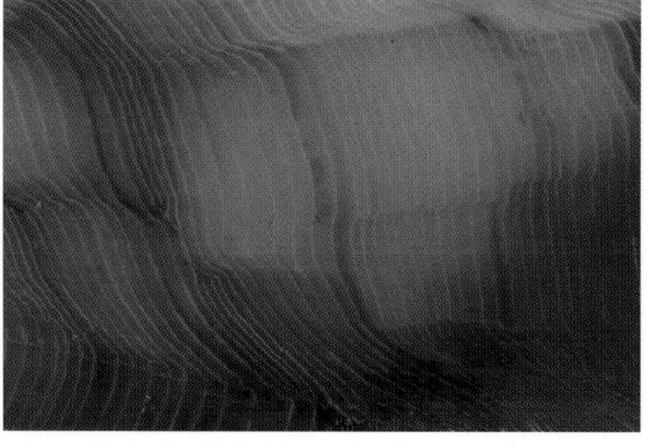

hydrothermal veins; secondary in the weathering zone of different types of rocks. It is often associated with chalcedony. Coatings and stalactitic aggregates of hyalite up to 50 mm (2 in) thick known from Cerritos, Mexico; Valec, Czech Republic; Klamath Falls, Oregon, USA. Milky opal occurs in Dubník, Slovakia; Smrcek, Czech Republic and many other localities. The most famous locality of fire opal is Zimapan, Hidalgo, Mexico. Precious opal comes from many localities in Australia, e.g. Baracoo River, Queensland; Coober Pedy, Southern Australia; and White Cliffs, New South Wales, where it forms rich aggregates and veinlets in sandstones; classic locality is Dubník, Slovakia, where it occurs in ande-

Dendritic opal, 56 mm, Křemže, Czech Republic

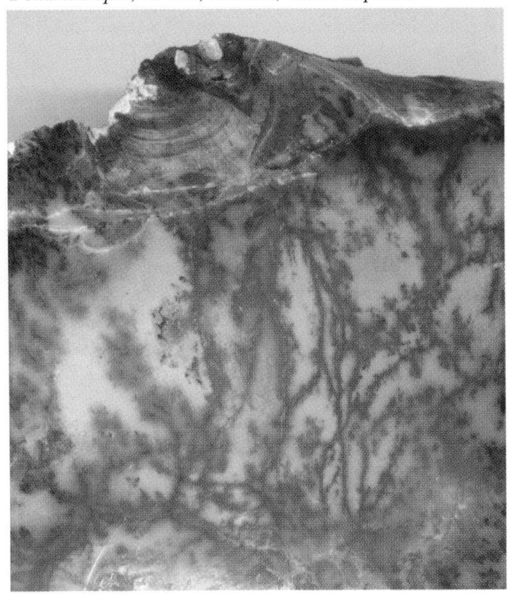

Fire opal, 20 mm, Mexico

sites and was probably mined already by ancient Romans. Beautiful precious opals come also from the Virgin Valley, Nevada, USA. Petrified trees, known as wooden opal, reach lengths of several meters in the Petrified Forest National Park, Holbrook, Arizona, USA; and in L'ubietová and Povrazník, Slovakia. White geysirite is known mainly from the hot springs in Iceland; Yellowstone National Park, Wyoming, USA; and New Zealand.

Application: colored opal varieties, primarily precious opal and fire opal are cut as gemstones, diatomite in chemical industry.

Precious opal, 55 mm, Opal Butte, U.S.A.

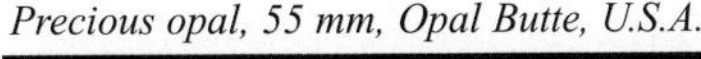

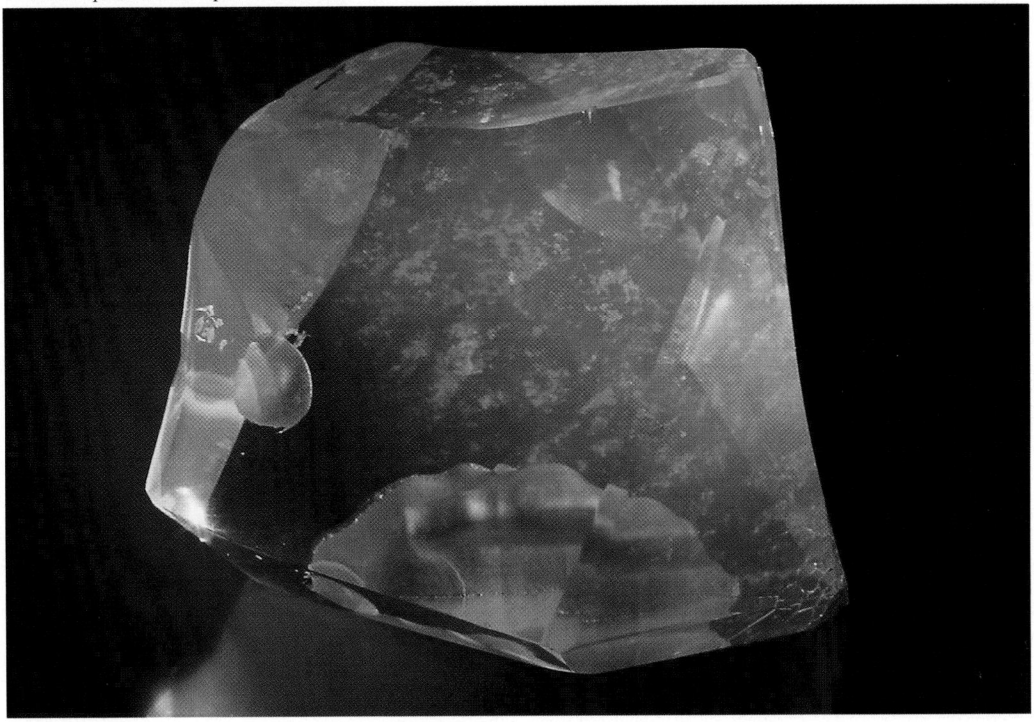

Melanophlogite, 80 mm, Fortullino, Italy

Melanophlogite

SiO_2

CUBIC ●

Properties: C – colorless, white; S – colorless; L – vitreous; D – transparent; DE – 2; H – 6.5; CL – none; M – pseudo-cubic crystals, spherical aggregates.
Origin and occurrence: Hydrothermal, associated with sulfur and other minerals. It was originally described from the sulfur deposit in Racalmuto, Sicily, Italy as crystals up to 4 mm (5/32 in) in size. It is also known from Chvaletice, Czech Republic, where it forms crystals up to 2 mm (1/16 in) in size, in Alpine-type veins.

Rutile

TiO_2

TETRAGONAL ● ● ● ●

Properties: C – red-brown, red, brown, yellowish, black; S – light brown; L – metallic to adamantine; D – transparent to translucent; DE – 4.2; H – 6-6.5; CL – good; F – conchoidal to uneven; M – short prismatic, striated crystals, common twins, acicular crystals, granular, massive.
Origin and occurrence: Magmatic and metamorphic, also in placers, together with monazite-(Ce), topaz, beryl, quartz and other minerals. The largest crystals up to 150 mm (6 in) in size come from the Mount Graves, Georgia, USA. Beautiful epitaxial intergrowths with hematite occur in Cavradischlucht, St. Gotthard, Switzerland and in Ibitiara, Bahía, Brazil. It is common as inclusions in smoky citrine (quartz) crystals from Ibitiara, Bahía and Itabira, Minas Gerais, Brazil. Knee-like crystal twins up to 70 (2¾ in) cm in size were found in the vicinity of Golcuv Jeníkov and Sobeslav, Czech Republic.
Application: Ti ore.

Rutile, 21 mm, Ibitiara, Brazil

Rutile, 48 mm, Bahía, Brazil

Cassiterite, 110 mm, Yunnan, China

Cassiterite

SnO_2

TETRAGONAL ● ● ● ●

Properties: C – colorless, brown, black; S – white, grayish, brown; L – metallic to adamantine, dull; D – transparent to opaque; DE – 6.3-7.2; H – 6-7; CL – imperfect; F – conchoidal to uneven; M – dipyramidal and short prismatic crystals, multiple twins, granular, massive.
Origin and occurrence: Magmatic in pegmatites, hydrothermal in high-temperature deposits, metamorphic and in placers, together with wolframite, topaz and other minerals. Crystal twins up to 150 mm (6 in) in size come from Horní Slavkov, Czech Republic. Crystals of similar size were also found in Panasqueira, Portugal. Fine twins up to 80 mm ($3\frac{1}{8}$ in) were found in Rossarden, Tasmania, Australia. Crystals up to 70 mm ($2\frac{3}{4}$ in) in size were found in Llallagua, colorless and transparent crystals up to 50 mm (2 in) in size in Viloco, Bolivia. Crystals up to 110 mm ($4\frac{5}{16}$ in) occurred lately in Tenkergin, Chukotka, Russia. Crystals up to 130 mm ($5\frac{2}{16}$ in) in size are known from pegmatites in Minas Gerais, Brazil (Fazenda do Funil). New finds of shiny crystals up to 100 mm (4 in) long were made in Hunan and Yunnan provinces, China.
Application: Sn ore.

Plattnerite

PbO_2

TETRAGONAL ● ●

Properties: C – black; S – brown; L – metallic to adamantine; D – opaque; DE – 9.6; H – 5.5; CL – none; M – acicular crystals, botryoidal aggregates, massive.
Origin and occurrence: Secondary, as a result of the oxidation of other Pb minerals, together with pyromorphite, hemimorphite and other minerals. Fine crystals come from Mina Ojuela, Mapimi, Durango, Mexico and from the Blanchard mine, New Mexico, USA. Botryoidal aggregates, weighing up to 100 kg (220 lb), were found in the Morning mine, Mullan, Idaho, USA.

Plattnerite, 150 mm, Mapimi, Mexico

Pyrolusite, 88 mm, Baraga, U.S.A.

Pyrolusite

MnO_2

TETRAGONAL ● ● ● ●

Properties: C – black, steel-gray; S – black; L – metallic to dull; D – opaque; DE – 5.1; H – 6-6.5; CL – perfect; F – uneven; M – prismatic to acicular striated crystals, stalactitic and botryoidal aggregates, granular, massive.

Origin and occurrence: Secondary, as a result of alteration of manganite and other primary Mn minerals, also hydrothermal. Crystals up to 20 mm ($^{25}/_{32}$ in) long come from Horní Blatná, Czech Republic. Radial shiny aggregates were found in Öhrenstock, Germany. It occurred in Ilfeld, Germany, too. Large sedimentary Mn deposits, where pyrolusite is the main constituent, are known near Chiaturi, Georgia or near Nikopol, Ukraine. Crystals are reported also from Tsumeb, Namibia and Hotazel, South Africa.

Application: important Mn ore.

Hollandite

$Ba(Mn^{4+},Mn^{2+})_8O_{16}$

MONOCLINIC ● ● ●

Properties: C – gray-black; S – black; L – submetallic; D – opaque; DE – 5; H – 6; CL – good; M – short prismatic crystals, racemous and columnar aggregates.

Origin and occurrence: Metamorphic in Mn deposits with braunite, scheelite and other minerals, also secondary. Crystals up to 5 mm ($^{3}/_{16}$ in) long come from the Bradshaw mountains, Arizona, USA. It is common in the metamorphosed Mn deposits in Ultevis, Sweden; in Nagpur and Balaghat, India.

Coronadite

$Pb(Mn^{4+},Mn^{2+})_8O_{16}$

TETRAGONAL ● ● ●

Properties: C – black, black-gray; S – brown-black; L – submetallic to dull; D – opaque; DE – 5.4; H – 4.5-5.5; M – botryoidal crusts with fibrous structure, massive.

Origin and occurrence: Secondary in the oxidation zone of Mn deposits. Spherical aggregates up to 100 mm (4 in) in diameter come from the Bou Tazoult mine, Imini, Morocco. Small crystals were found in the Beltana mine, Southern Australia, Australia and in the Silver Bill mine, Arizona, USA.

Todorokite

$(Mn^{2+},Ca,Mg)Mn^{4+}{}_3O_7 . H_2O$

MONOCLINIC ● ● ●

Properties: C – gray, brown-black, black; S – brown;

Hollandite, 90 mm, Ultevis, Sweden

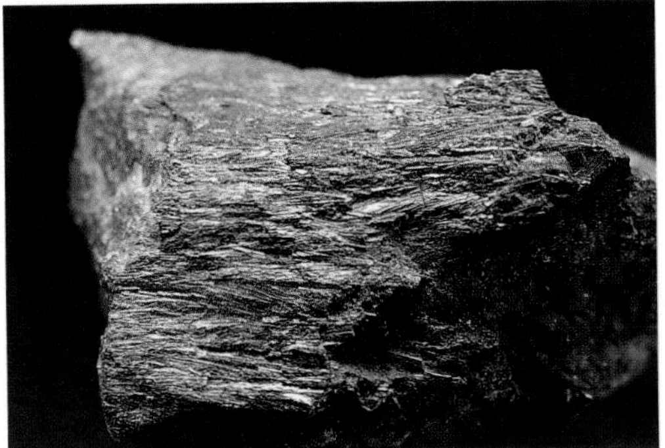

Coronadite, 80 mm, Imini, Morocco

L – metallic to dull; D – opaque; DE – 3.5-3.8; H – 1; CL – perfect; M – platy crystals, stalactitic and nodular aggregates.
Origin and occurrence: Secondary mineral, as a result of the oxidation of Mn minerals. Crystals are reported from several mines in the Kalahari region, South Africa (Hotazel, Smart). It was originally described from the Todoroki mine, Hokkaido, Japan. It is an important constituent of oceanic Mn concretions.

Todorokite, 60 mm, Kamogun, Japan

Ferrotapiolite

$FeTa_2O_6$

TETRAGONAL ● ● ●

Properties: C – black, brown; V – red-brown; L – submetallic, adamantine, resinous; D – opaque; DE – 7-7.8; H – 6-6.5; CL – none; F – uneven to conchoidal; M – dipyramidal and short prismatic crystals, massive.
Origin and occurrence: Magmatic in pegmatites, together with manganotantalite, microlite, cassiterite and other minerals. Crystals up to 40 mm ($1^{9}/_{16}$ in) in size are known from the vicinity of Governador Valadares, Minas Gerais, Brazil. A crystal 120 mm ($4^{7}/_{16}$ in) long has been described from Angarf, Morocco. Short prismatic crystals come from pegmatites near Topsham and Paris, Maine, USA.

Ferrotapiolite, 80 mm, Maršíkov, Czech Republic

Ilmenorutile

$(Ti,Nb,Fe)O_2$

TETRAGONAL ● ● ●

Properties: C – black; S – brown; L – submetallic; D – opaque; DE – 4.2; H – 6-6.5; CL – good; F – conchoidal to uneven; M – prismatic crystals, granular.
Origin and occurrence: Magmatic in pegmatites, together with schorl, zircon, fluorapatite and other minerals. Prisms several cm long come from Údraz near Písek, Czech Republic. It is also known from the vicinity of Miass, Ural mountains, Russia and Evje, Norway.

Ilmenorutile, 10mm x, Písek, Czech Republic

Anatase
TiO_2

TETRAGONAL ● ● ●

Properties: C – black-gray, brown, red-brown, blue, rare colorless; S – white; L – submetallic to adamantine; D – transparent to opaque; DE – 3.8-4; H – 5.5-6; CL – perfect; F – conchoidal; M – dipyramidal and tabular crystals.
Origin and occurrence: Hydrothermal in the Alpine-type veins, associated with brookite and quartz, also sedimentary and metamorphic. Beautiful crystals up to 50 mm (2 in) long were found in Alpa Lercheltini, Binntal, Switzerland. Famous black-blue crystals reaching up to 30 mm (1 3/16 in) come from Matskorhae, Hardangervidda, Norway. Crystals up to 15 mm (1 9/32 in) were recently found in Dodo, Polar Ural, Russia. Crystals up to 30 mm (1 3/16 in) in size occurred in the Old Lot and Vulcan mines, Colorado, USA.

Tellurite
TeO_2

ORTHORHOMBIC ● ●

Properties: C – yellow, yellow-orange; S – yellowish; L – adamantine; D – transparent; DE – 5.8; H – 2; CL – perfect; M – acicular and thin tabular crystals, radial aggregates, pulverulent.
Origin and occurrence: Secondary, resulting from the oxidation of AuBTe ores. Crystals up to 10 mm (3/8 in) long occurred in the Kawazu and Susaki mines, Japan. Beautiful specimens were found in Moctezuma, Sonora, Mexico; it also comes from Cripple Creek, Colorado, USA.

Brookite
TiO_2

ORTHORHOMBIC ● ● ●

Properties: C – light to dark brown, yellow-brown, black; S – white to gray; L – submetallic to adamantine; D – transparent to translucent, opaque; DE – 4.1; H – 5.5-6; CL – imperfect; F – conchoidal to uneven; M – tabular, dipyramidal, long and short prismatic crystals.
Origin and occurrence: Hydrothermal along the fissures of the Alpine-type veins and in granitic and alkaline pegmatites; it occurs as pseudo-morphs after titanite and ilmenite; also in sedimentary rocks. Perfect tabular crystals up to 50 mm (2 in) in size found in Rieder Tobel, Switzerland; Magnet Cove, Arkansas, USA; and Passo di Viza, Italy. New finds of crystals up to 50 mm in size made in Dodo, Polar Ural, Russia.

Anatase, 16 mm, Hardangervidda, Norway

Brookite, 50 mm, Puiva, Polar Urals, Russia

Tellurite, 1 mm x, Fata Baii, Romania

Ferberite

$FeWO_4$

MONOCLINIC ● ● ●

Properties: C – black; S – brown-black to black; L – submetallic; D – opaque; DE – 7.5; H – 4-4.5; CL – perfect; F – uneven; M – short prismatic to tabular crystals, granular, massive.
Origin and occurrence: Hydrothermal in high- to medium-temperature ore veins, in greisens and skarns; rare magmatic in granitic pegmatites and granites; it also occurs in placers. It is usually associated with cassiterite, scheelite, sulfides and quartz. Perfect tabular crystals up to 120 mm (4 7/16 in) in size found in the Quartz Creek mine, Colorado, USA; also in Cínovec, Czech Republic; Panasqueira, Portugal; Ehrenfriedersdorf, Germany; and Potosí, Bolivia.
Application: W ore.

Hübnerite

$MnWO_4$

MONOCLINIC ● ● ●

Properties: C – yellow-brown, red-brown, black; S – yellow-brown to black-gray; L – submetallic; D – translucent to opaque; DE – 7.2; H – 4-4.5; CL – perfect; F – uneven; M – short prismatic to tabular crystals, granular, massive.
Origin and occurrence: Hydrothermal in high- to medium-temperature ore veins, in greisens; rarely magmatic in granitic pegmatites; also in placers. Perfect tabular to short prismatic crystals up to 25 cm (9 13/16 in) in size, come from the Huayllapon mine, Pasto Bueno, Peru; also from Baia Sprie, Romania; Kara-Oba, Kazakhstan; the Sweet Home mine, Alma; and Silverton, Colorado, USA.
Application: W ore.

Ferberite, 68 mm, Mundo Nuovo, Peru

Hübnerite, 39 mm, Silverton, U.S.A.

Titanowodginite, 20 mm, Tanco, Canada

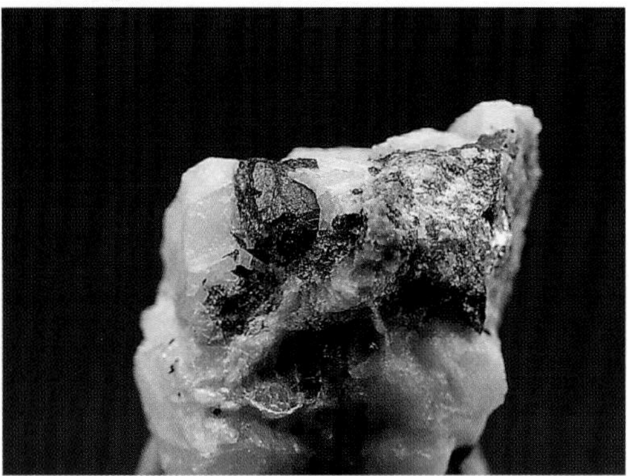

Titanowodginite

$MnTiTa_2O_8$

ORTHORHOMBIC ●

Properties: C – dark brown, black; S – dark brown; L – submetallic; D – translucent to opaque; DE – 6.9; H – 5.5; CL – imperfect; F – uneven; M – dipyramidal crystals, granular.
Origin and occurrence: Magmatic in granitic pegmatites. Dipyramidal crystals up to 10 mm (3/8 in) long occur in the Tanco mine, Bernic Lake, Manitoba, Canada.
Application: Ta ore.

Ferrocolumbite, 90 mm, Middletown, U.S.A.

Ferrocolumbite

$FeNb_2O_6$

ORTHORHOMBIC ● ● ●

Properties: C – black, red-brown; S – red-brown to black; L – submetallic; D – translucent to opaque; DE – 5.2; H – 6; CL – good; F – uneven to conchoidal; M – long to short prismatic and tabular crystals, granular, massive.
Origin and occurrence: Magmatic in granitic pegmatites and granites; rare hydrothermal in high-temperature ore veins and in greisens; also in placers. Tabular crystals up to 1 m (39 3/8 in) in size occur in granitic pegmatites near Custer and Keystone, South Dakota, USA; in Malakialina, Madagascar; Ichikawa, Japan; masses, weighing up to 270 kg (594 lb) come from the Meyers quarry, Colorado, USA.
Application: Nb ore.

Manganotantalite

$MnTa_2O_6$

ORTHORHOMBIC ● ● ●

Properties: C – red, red-brown, black-brown, black; S – dark red to black; L – submetallic; D – translucent to opaque; DE – 8.O; H – 6-6.5; CL – good; F – uneven

Manganotantalite, 19 mm, Nuristan, Afghanistan

Euxenite-(Y), 40 mm, Ambatofotsy, Madagascar

Aeschynite-(Ce), 45 mm, Hitterö, Norway

to conchoidal; M – prismatic to tabular crystals, granular, massive.
Origin and occurrence: Magmatic in granitic pegmatites and granites; also in placers. Crystals up to 100 mm (4 in) in size come from Li-bearing pegmatites in the Tanco Mine, Bernic Lake, Manitoba, Canada; also from Sao Jose da Safira, Minas Gerais, Brazil.
Application: Ta ore.

Euxenite-(Y)

(Y,Ce,U,Th) $(Nb,Ta,Ti)_2O_6$

ORTHORHOMBIC ● ●

Properties: C – black with brownish and green hues; S – gray, yellowish, brownish; L – submetallic, resinous; D – translucent to opaque; DE – 4.6; H – 6; CL – none; F – conchoidal; M – tabular crystals, granular, massive; R – locally weakly radioactive, commonly metamict.
Origin and occurrence: Magmatic in granitic and alkaline pegmatites; also in placers. Typically associated with monazite-(Ce), zircon, ilmenite and other oxides of rare earth elements. Crystals up to 150 mm (6 in) in size are known from Kragerö and Hitterö, Norway; from Ankazobé, Madagascar.

Aeschynite-(Ce)

(Ce,Ca) $(Ti,Nb)_2O_6$

ORTHORHOMBIC ● ●

Properties: C – black, red-brown, yellow; S – red-yellow; L – vitreous, resinous, adamantine; D – translucent to opaque; DE – 5.0; H – 5.5; CL – none; F – conchoidal; M – prismatic and tabular crystals, granular, massive; R – locally weakly radioactive, commonly metamict.
Origin and occurrence: Magmatic in alkaline and granitic pegmatites and carbonatites. It is associated with zircon and oxides of rare earth elements. Crystals up to 190 mm ($7^{8}/_{16}$ in) long occur in Quadeville, Ontario, Canada; other localities are Kragerö, Norway; and Trout Creek Pass, Colorado, USA.

Stibiotantalite

$SbTaO_4$

ORTHORHOMBIC ● ●

Properties: C – yellow, yellow-brown, red-brown, yellow-green; S – yellow-brown; L – submetallic, vitreous, resinous; D – transparent to translucent; DE – 7.5; H – 5-5.5; CL – good; F – uneven to conchoidal; M – prismatic, tabular and dipyramidal crystals, granular, massive.
Origin and occurrence: Magmatic in granitic pegmatites; also in placers. Crystals up to 120 mm ($4^{7}/_{16}$ in) in size occur in Muiane, Alto Ligonha, Mozambique; also found in the Little Three mine, Ramona; and the Himalaya mine, Mesa Grande, California, USA. Also known from Greenbushes, Western Australia.
Application: Ta ore.

Stibiotantalite, 10 mm grain, Dobrá Voda, Czech Republic

Uraninite
UO_2

CUBIC ● ● ● ●

Properties: C – black, black-brown, black-gray; S – black, black-brown to greenish; L – submetallic, greasy, earthy; D – opaque; DE – 7.5-10,6; H – 5-6. earthy aggregates 3; CL – imperfect; F – uneven to conchoidal; M – cubic crystals, botryoidal aggregates, granular, massive; R – strong radioactive.
Origin and occurrence: Mainly hydrothermal in ore veins, skarns; magmatic in granitic pegmatites; in sedimentary rocks; also in placers. Usually associated with other U minerals, e.g. coffinite and secondary alteration products, mainly U micas. Perfect crystals up to 100 mm (4 in) in size and weighing up to 2.5 kg (5 lb 8 oz) come from the Fissure mine, Wilberforce, Ontario, Canada, where they occur in simple pegmatites, cross-cutting marbles. Crystals are also known from Dieresis, Spain and Shinkolobwe, Zair. Rich botryoidal aggregates were found in Jáchymov and Slavkovice, Czech Republic; in Bois-Noirs and Margnac, France.
Application: U ore.

Uraninite, 8 mm x, Portland, U.S.A.

Gibbsite, 80 mm, Gamba, Brazil

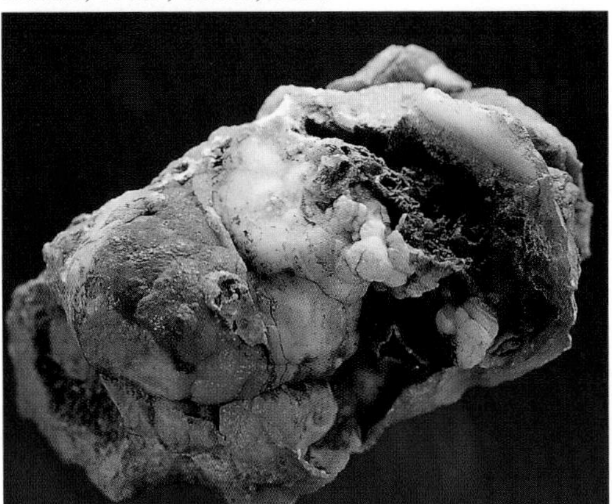

Brucite, 30 mm, Azbest, Russia

Gibbsite
$Al(OH)_3$

MONOCLINIC ● ● ●

Properties: C – colorless, gray, white, greenish; S – white; L – vitreous, pearly; D – transparent to translucent; DE – 2.4; H – 2.5-3.5; CL – perfect; F – uneven; M – tabular crystals, lamellar and earthy aggregates, coatings and stalactitic films, granular, massive.
Origin and occurrence: Hydrothermal as a product of alteration Al-rich rocks; secondary in the oxidation associated with goethite; metamorphic in weakly metamorphosed Al-rich rocks, typically with diaspore; a constituent of bauxites. Tabular crystals up to 100 mm (4 in) in size were found in Zlatoust, Ural mountains, Russia; also in Villa Rica, Minas Gerais, Brazil.

Brucite
$Mg(OH)_2$

TRIGONAL ● ● ●

Properties: C – colorless, gray, white, bluish, blue, yellow, brown; S – white; L – vitreous, pearly; D – transparent to translucent; DE – 2.4; H – 2.5; CL – perfect; F – uneven; M – tabular crystals, foliated, acicular and earthy aggregates, granular, massive.
Origin and occurrence: Hydrothermal in veins in serpentinites or dolomitic marbles, a product of periclase alteration; rare metamorphic in skarns and marbles. Perfect crystals up to 18 cm in size, come from the Low's mine, Pennsylvania and the Tilly Foster mine, New York, USA; also known from Asbestos, Quebec, Canada; Predazzo, the Alps, Italy; blue crystals up to 50 mm (2 in) in size were found in the Bazhenovskoye deposit, Azbest, Ural mountains, Russia.

Diaspore
AlO(OH)

ORTHORHOMBIC ● ● ●

Properties: C – colorless, gray, white, greenish, yellowish, pink, purplish; S – white; L – vitreous, pearly; D – transparent to translucent; DE – 3.4; H – 6.5-7; CL – perfect; F – conchoidal; M – tabular crystals, foliated aggregates, stalactitic films, granular, massive.
Origin and occurrence: Hydrothermal as a product of alteration Al-rich minerals, e.g. andalusite, typically with pyrophyllite and corundum; metamorphic in Al-rich rocks; a constituent of bauxites. Tabular crystals up to 120 mm (4 7/16 in) in size come from Menderess, Turkey; also from Naxos, Greece; and Chester, Massachusetts, USA.

Diaspore, 35 mm, Milas, Turkey

Goethite, 77 mm, Santa Eulalia, Mexico

Goethite, 60 mm, Příbram, Czech Republic

Goethite
$Fe^{3+}O(OH)$

ORTHORHOMBIC ● ● ● ● ●

Varieties: velvet ore

Properties: C – black-brown, yellow-brown, brown; S – yellow-brown; L – submetallic, metallic, silky, earthy; D – translucent to opaque; DE – 4.3; H – 5-5.5; CL – perfect; F – uneven to conchoidal; M – acicular to prismatic crystals, botryoidal aggregates, commonly with radial structure, coatings and stalactitic films, earthy, granular, massive.
Origin and occurrence: Secondary as one of the most common minerals of the oxidation zone of ore deposits, it forms a significant part of limonite; hydrothermal in ore veins, in cavities in pegmatites and volcanic rocks. It forms pseudo-morphs after pyrite and other Fe sulfides. Rich botryoidal aggregates of velvet ore with a velvety surface come from Príbram, Czech Republic; acicular crystals up to 50 mm (2 in) long are known from Bottalack and Redruth, Cornwall, UK; it also occurs in Siegen and Horhausen, Germany; and in Florissant, Colorado, USA.
Application: Fe ore.

Lepidocrocite, 30 mm, Rudabánya, Hungary

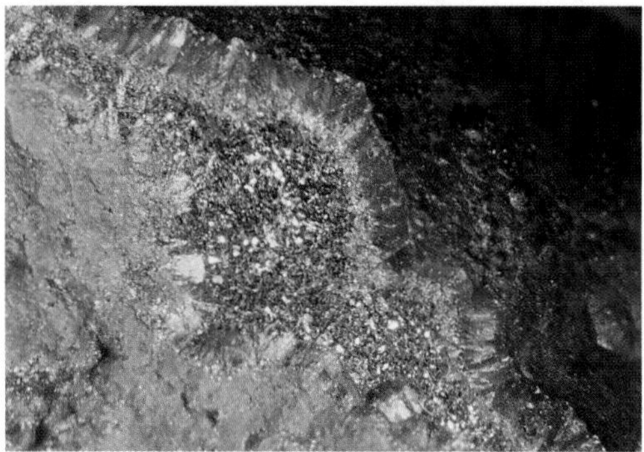

Manganite
$Mn^{3+}O(OH)$

MONOCLINIC ● ● ● ●

Properties: C – black to black-gray; S – red-brown to black; L – submetallic to dull; D – opaque; DE – 4.3; H – 4; CL – perfect; F – uneven to conchoidal; M – long to short prismatic crystals, acicular and earthy aggregates, concretions, granular, massive.
Origin and occurrence: Hydrothermal in low-temperature ore veins, together with quartz; secondary in the oxidation zone of ore deposits; sedimentary and rare metamorphic in Mn-rich rocks. Druses of black crystals up to 40 mm ($1^{9}/_{16}$ in) long come from the classic locality Ilfeld, Germany; it also occurs in Ohrenstock and Ilmenau, Germany; in Nikopol, Ukraine; in Sterling Hill, New Jersey, USA; and in the N'Chwaning No. 2 mine, Kuruman, South Africa.
Application: Mn ore.

Lepidocrocite
$Fe^{3+}O(OH)$

ORTHORHOMBIC ● ● ●

Properties: C – dark red to red-brown; S – orange to brick-red; L – submetallic, adamantine to silky; D – transparent to opaque; DE – 4.0; H – 5; CL – perfect; F – uneven to conchoidal; M – tabular to short prismatic crystals, acicular, bladed and earthy aggregates, concretions, granular, massive.
Origin and occurrence: Secondary in the oxidation zone of ore deposits, overgrown on botryoidal goethite. It occurs together with goethite as a constituent of limonite, its tabular crystals and their aggregates are known from Herdorf, Germany and Rancié, France.

Manganite, 85 mm, Ilfeld, Germany

Lithiophorite

$(Al,Li)Mn^{4+}O_2(OH)_2$

MONOCLINIC ● ● ●

Properties: C – black, commonly with bluish tint; S – black-gray to black; L – metallic to dull; D – opaque; DE – 3.3; H – 3; CL – perfect; F – uneven; M – scaly crystals, botryoidal and earthy aggregates, coatings, granular, massive.

Origin and occurrence: Secondary in the oxidation zone of ore deposits and along the cracks in sedimentary rocks. Botryoidal aggregates occur in Schneeberg, Germany; Jivina and Zajecov, Czech Republic; and Miyazaki, Japan.

Lithiphorite, 60 mm, Rangersdorf, Germany

Curite

$Pb_2U_5O_{17} . 4 H_2O$

ORTHORHOMBIC ● ●

Properties: C – dark orange to red-orange; S – light orange; L – adamantine to earthy; D – transparent to translucent; DE – 7.4; H – 4-5; CL – good; F – uneven; M – acicular crystals, earthy aggregates, coatings, massive; R – strong radioactive.

Origin and occurrence: Secondary in the oxidation zone of U deposits, associated with other secondary U minerals, e.g. torbernite, kasolite and uranophane. Rich aggregates were found in Shinkolobwe, Zaire; also known from La Crouzille, France; and South Alligator, Northern Territory, Australia.

Curite, 30 mm, Shinkolobwe, Zair

5. Carbonates

Magnesite

$MgCO_3$

TRIGONAL ● ● ● ●

Properties: C – colorless, white, yellowish, brownish, black; S – white; L – vitreous to dull; D – transparent; DE – 3.1; H – 4; CL – perfect; F – conchoidal; M – rhombohedral and prismatic crystals, massive cleavable aggregates, earthy; LU – occasionally blue or green.
Origin and occurrence: Rarely magmatic, mainly hydrothermal metasomatic and metamorphic. The largest crystals are known from Brumado, Bahía, Brazil, reaching up to 1 m (39 3/8 in) in size, embedded in metamorphosed dolomites. Crystals in cavities in the same locality are up to 50 mm (2 in) in size. Also crystals up to 50 mm (2 in) found in the Eugui quarries, Spain. Crystals up to 10 mm (3/8 in) across come also from Val Malenco, Italy. It prevails as massive aggregates, forming huge deposits, like Veitsch, Austria; Liao-Tung, China. Many deposits are located in Slovakia (Jelšavská Dúbrava, Hnúšta).
Application: heat-resistant material.

Calcite, 40 mm, Houghton Co., U.S.A.
Magnesite, 148 mm, Brumado, Brazil

Smithsonite, 32 mm, Tsumeb, Namibia

Smithsonite

$ZnCO_3$

TRIGONAL ● ● ●

Properties: C – white, gray, green, pink, blue; S – white; L – vitreous to pearly; D – transparent to translucent; DE – 4.4; H – 4-5; CL – perfect; F – conchoidal to uneven; M – rhombohedral crystals, botryoidal and stalactitic aggregates, massive; LU – sometimes green or blue.
Origin and occurrence: Supergene, as a result of oxidation of the primary Zn ores, associated with other supergene Pb minerals. The largest yellow scalenohedra crystals up to 40 mm (1 9/16 in) in size come from Broken Hill, New South Wales, Australia. Pink crystals, up to 30 mm (1 3/16 in) long occurred in Tsumeb, Namibia. World famous blue-green botryoidal crusts up to 100 mm (4 in) thick found in the Kelly Mine, Magdalena, New Mexico, USA. Nice aggregates and banded stalactites discovered in Monte Poni, Sardinia, Italy.

Smithsonite, 40 mm, New Mexico, U.S.A.

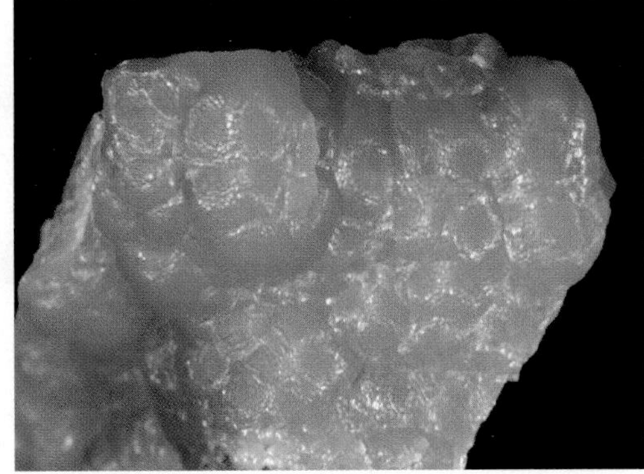

Siderite, 26 mm, Governador Valadares, Brazil

Siderite

$FeCO_3$

TRIGONAL ● ● ● ●

Properties: C – yellow, brown, black; S – yellowish-white; L – vitreous; D – translucent; DE – 4; H-4; CL – perfect; F – uneven to conchoidal; M – rhombohedral crystals, granular, massive.

Origin and occurrence: Hydrothermal in medium- and low-temperature deposits, sedimentary. Crystals up to 40 cm long found in Mont St.-Hilaire, Quebec, Canada. Crystals up to 100 mm (4 in) in size come from Panasqueira, Portugal. Crystals up to 30 mm (1³/₁₆ in) in size occurred in Neudorf, Germany. Rhombs up to 20 mm (²⁵/₃₂ in) also found in Příbram, Czech Republic. Pseudo-morphs of goethite after siderite up to 70 mm (2¾ in) across described from Pikes Peak, Colorado, USA. Deposits of massive siderite in Erzberg and Hüttenberg, Austria yielded crystals up to 50 mm (2 in) long. Fine crystals reported from Tavistock, Devon and Redruth, Cornwall, UK.

Application: important Fe ore.

Sfaerocobaltite

$CoCO_3$

TRIGONAL ● ● ●

Properties: C – pink, gray, brown; S – red; L – vitreous; D – transparent to translucent; DE – 4.1; H – 4; CL – perfect; M – scalenohedral and rhombohedral crystals, radial aggregates, massive.

Origin and occurrence: Secondary, as a product of the oxidation of primary Co minerals. The best specimens, with crystals up to 30 mm (1³/₁₆ in) long, come from Zaire (Musonoi; Kakanda). Crystals up to 10 mm (4 in) long known from Bou Azzer, Morocco.

Sphaerocobaltite, 16 mm, Bou Azzer, Morocco

Rhodochrosite, 117 mm, Sweet Home Mine, U.S.A.

Rhodochrosite

$MnCO_3$

TRIGONAL ● ● ● ●

Properties: C – white, pink, red, brown, locally black coatings on crystals; S – white; L – vitreous; D – transparent to translucent; DE – 3.6; H – 3.5-4; CL – perfect; F – conchoidal to uneven; M – rhombohedral and scalenohedral crystals, hemispherical and botryoidal aggregates, granular, massive.

Origin and occurrence: Only rare in pegmatites, hydrothermal in medium- and low-temperature veins, sedimentary and metamorphic. The most beautiful crystals come from the Sweet Home mine, Alma, Colorado, USA, where rhombs up to 150 mm (6 in) in size occur, associated with purple fluorite, hübnerite, tetrahedrite and other minerals. Beautiful dark red scalehohedra, up to 100 mm (4 in) found in the N'Chwaning No.1 and 2 mines, Kuruman, South Africa. Pink rhombs up to 80 mm (3 1/8 in) come from Silverton, Colorado, USA. Nice pink hemispheres and botryoidal aggregates are known from Cavnic and Baia Sprie, Romania. Similar specimens occurred in Huaron, Peru. Pink banded crusts and stalactites were found in the Mina Capillitas, Catamarca, Argentina. Pink rhombs, associated with bertrandite, were lately found in Kounrad, Kazakhstan.

Rhodochrosite, 70 mm, Mina Capillitas, Argentina

Calcite, 150 mm, Elmwood Mine, U.S.A.

Calcite
$CaCO_3$

TRIGONAL ● ● ● ● ●

Properties: C – colorless, white, gray, yellow, brown, pink, red, blue, green, black; S – white; L – vitreous to pearly; D – transparent to opaque; DE – 2.7; H – 3; CL – perfect; F – conchoidal; M – crystals of various habit, concretions, stalactites, oolitic aggregates, granular, massive; LU – sometimes red to orange.

Calcite, 30 mm, Tunguzka, Russia

Origin and occurrence: One of the most common minerals, resulting from a wide range of conditions, it is magmatic, hydrothermal, sedimentary, metamorphic and secondary, it occurs in various parageneses. Large crystals found in many localities throughout the world. Pinkish and yellow crystals over 500 mm (20 in) long come from Joplin, Missouri and from the Elmwood mine, Tennessee, USA. Beautiful calcite crystals, crystallographically of very complex habits, found in Dalnegorsk, Russia. Nice calcite specimens occurred also in Mexico (Naica, Chihuahua; Charcas, San Luis Potosí). European localities, like Príbram, Czech Republic; St. Andreasberg, Germany; Kongsberg, Norway are famous by their calcites, too. Wine yellow, complicated crystals are known from the Sarbayskoye deposit in Rudnyi, Kazakhstan. Clear cleavable aggregates of the birefringent calcite (so called Iceland spar) were found in basalt cavities in Helgustadir, Iceland. The largest of them reached up to 6 x 2 m (20 x 6 ft 6 in) in size. Very nice scalenohedra up to 80 mm ($3\frac{1}{8}$ in) long with copper inclusions occur in Keweenaw Peninsula, Michigan, USA. Perfect scalenohedra and their twins up to 100 mm (4 in) long are known from Egremont and Frizington, UK. Beautiful butterfly twins up to 80 mm ($3\frac{1}{8}$ in) recently reported from Guiyang, Hunan, China. Crystals of calcite with sand inclusions up to 100 mm (4 in) in size come from the vicinity of Fontainebleau, France.

Application: building industry, optical industry.

Dolomite, 33 mm x, Navarro, Spain

Dolomite

$CaMg(CO_3)_2$

TRIGONAL ● ● ● ● ●

Properties: C – gray-white, pink, red, green, brown, black; S – white; L – vitreous to pearly; D – transparent to translucent, DE – 2.9; H – 3.5; CL – perfect; F – conchoidal; M – rhombohedral crystals, massive.
Origin and occurrence: Magmatic in pegmatites, hydrothermal, metasomatic, sedimentary and metamorphic, together with siderite, magnesite, calcite and other minerals. Crystals up to 100 mm (4 in) long found in Brumado, Bahía, Brazil. Fine crystals also occurred in Banská Štiavnica, Slovakia and in Cavnic, Romania. Crystals up to 200 mm ($7\frac{7}{8}$ in) in size come from Eugui, Spain. Crystals up to 150 mm (6 in) long found in cavities in dolomitic rocks in Lengenbach, Binntal, Switzerland. Crystals several cm long known from Jáchymov, Czech Republic. Large accumulations of massive dolomite are common in magnesite deposits.
Application: metallurgy, heat-resistant material.

Ankerite, 20 mm xx, Roudny, Czech Republic

Ankerite

$CaFe(CO_3)_2$

TRIGONAL ● ● ● ●

Properties: C – white, yellowish, brown-yellow; S – white; L – vitreous to pearly; D – translucent; DE – 3; H – 3.5-4; CL – perfect; F – conchoidal; M – rhombohedral crystals, granular.
Origin and occurrence: Hydrothermal in medium- and low-temperature veins, also sedimentary and metamorphic, together with siderite and other minerals. Crystals up to 50 mm (2 in) occurred in the Tui mine, New Zealand. Brown rhombs up to 40 mm ($1\frac{9}{16}$ in) in size known from Gilman, Colorado, USA. Yellowish crystals up to 10 mm ($\frac{3}{8}$ in) come from concretions near Kladno, Czech Republic. Massive aggregates are common in metasomatic deposits of siderite (e.g. Nizná Slaná, Slovakia).

Kutnohorite, 60 mm, Kutná Hora, Czech Republic

Huntite, 110 mm, Kokšín, Czech Republic

Kutnohorite

$CaMn(CO_3)_2$

TRIGONAL ● ● ●

Properties: C – white, gray, pink, yellowish; S – white; L – vitreous; D – translucent; DE – 3.1; H – 3.5-4; CL – perfect; F – conchoidal; M – poorly developed crystals, granular, massive.
Origin and occurrence: Hydrothermal and metamorphic, associated with ankerite, quartz and other minerals. Poorly developed crystals are known from Kutná Hora, Czech Republic. Small crystals several mm in size are described from Mont St.-Hilaire, Quebec, Canada. Large crystals occurred in Moncure, North Carolina, USA. Small gray-white crystals were found in Broken Hill, New South Wales, Australia.

Huntite

$CaMg(CO_3)_2$

TRIGONAL ● ●

Properties: C – white; S – white; L – earthy; D – opaque; DE – 2.7; H – 1.5; CL – none; F – conchoidal; M – fibrous aggregates, earthy.
Origin and occurrence: Secondary mineral resulting from the oxidation of dolomite, associated with magnesite and dolomite. Fine fibrous aggregates come

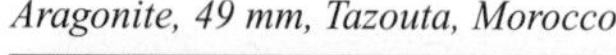

Aragonite, 49 mm, Tazouta, Morocco

Flos ferri, 110 mm, Erzberg, Austria

from Kokšín, Czech Republic. It also occurs in the Ala-Mar deposit, Nevada, USA. Massive aggregates are known from the Boquira mine, Bahía, Brazil.

Aragonite
$CaCO_3$

ORTHORHOMBIC ● ● ● ●

Varieties: flos ferri, hot-spring tufa, peastone, tarnowitzite.
Properties: C – colorless, white, yellow, reddish, greenish, purplish, bluish, gray; S – white; L – vitreous; D – transparent to opaque; DE – 3; H – 3.5-4.5; CL – imperfect; F – conchoidal; M – prismatic crystals, oolitic, banded, columnar and dendritic aggregates, massive; LU – locally weak cream yellow.
Origin and occurrence: Primary as a late hydrothermal mineral of high-temperature deposits more commonly secondary as a product of the oxidation of siderite and pyrite. It also results from precipitation of thermal springs, it is sedimentary and metamorphic. Fine white prismatic crystals up to 70 mm ($2\frac{3}{4}$ in) long found together with blue celestite crystals in Spania Dolina, Slovakia. Similar crystals come from sulfur deposits in Cianciano, Italy and Tarnobrzeg, Poland. Maybe the best aragonite crystals in the world are known from the gossan of the magnesite deposit in Podrecany, Slovakia, where druses of crystals up to 200 mm ($7\frac{7}{8}$ in) long were found. Wine yellow crystals up to 100 mm ($\frac{3}{8}$ in) long occurred in cavities of volcanic rocks in Horenec near Bílina, Czech Republic. Those crystals were the only gem rough, suitable for facetting, in the world. Very interesting copper pseudo-morphs after aragonite come from Corocoro, Bolivia. Banded and oolitic aggregates found in Karlovy Vary, Czech Republic. Dendritic aggregates from Erzberg, Austria and elsewhere are known as flos ferri.
Application: decorative stone.

Peastone, 35 mm, Karlovy Vary, Czech Republic

Strontianite, 34 mm, Hardin Co., U.S.A.

Strontianite

$SrCO_3$

ORTHORHOMBIC ● ● ●

Properties: C – colorless, gray, yellowish, greenish, reddish, brown; S – white; L – vitreous to resinous; D – transparent to translucent; DE – 3.8; H – 3.5; CL – perfect; F – conchoidal to uneven; M – prismatic and acicular crystals, columnar and fibrous aggregates, massive, earthy.
Origin and occurrence: Hydrothermal in low-temperature veins, in cavities of volcanic rocks, mainly sedimentary, together with calcite and zeolites. Crystals up to 80 mm (3 1/8 in) long occur near Bleiberg, Austria. Smaller crystals known from Strontian, Scotland. Crystals up to 50 mm long found in marls in Ahlen near Münster, Germany.

Witherite

$BaCO_3$

ORTHORHOMBIC ● ●

Properties: C – white, gray, yellowish; S – white; L – vitreous to greasy, D – translucent; DE – 4.3; H – 3.5; CL – good; F – uneven; M – pseudo-hexagonal dipyramidal crystals, fibrous and botryoidal aggregates, massive.
Origin and occurrence: Hydrothermal in low-temperature deposits with fluorite, barite, calcite and other minerals. Beautiful yellowish crystals up to 120 mm (4 11/16in) long come from the Minerva No.1 mine, Cave-in-Rock, Illinois, USA. Crystals, up to 70 mm (2¾ in) in size occurred in Hexham and Alston Moor, England, UK. Its pseudo-hexagonal crystals or botryoidal aggregates were very rare in Příbram, Czech Republic.

Cerussite

$PbCO_3$

ORTHORHOMBIC ● ● ●

Properties: C – colorless, whitish, yellow, black; S – white; L – greasy to adamantine; D – translucent; DE – 6.6; H – 3-3.5; CL – good; F – conchoidal; M –

Witherite, 20 mm x, Alston Moor, UK

Cerussite, 92 mm, Tsumeb, Namibia

prismatic and pyramidal crystals, common trillings and twins, granular, massive; LU – sometimes yellowish.
Origin and occurrence: Secondary mineral, resulting from the oxidation of galena and other Pb minerals, together with pyromorphite, vanadinite, barite and other minerals. The best specimens come from Tsumeb, Namibia, where trillings up to 200 mm ($7^{7}/_{8}$ in) in diameter occurred. Large twins also known from Broken Hill, New South Wales, Australia. Beautiful crystals up to 50 mm (2 in) in size reported from Mibladen, Morocco. Fine crystals up to 50 mm (2 in) long found in Stříbro, Czech Republic. Typical white acicular crystals, up to 60 mm ($2^{3}/_{8}$ in) long come from the Flux mine, Arizona, USA.

Barytocalcite
$BaCa(CO_3)_2$

MONOCLINIC ● ●

Properties: C – colorless, white, gray, yellowish; S – colorless; L – vitreous to resinous; D – transparent to translucent; DE – 3.7; H – 4; CL – perfect; F – conchoidal to uneven; M – prismatic, often striated crystals, massive; LU – light yellow.
Origin and occurrence: Hydrothermal in low-temperature veins, together with calcite, barite and other minerals. Crystals several cm in size and cleavable masses are known from Alston Moor, Cumbria, UK. Imperfect crystals about 10 mm ($^{3}/_{8}$ in) in size occurred in Stříbro, Czech Republic. It is also described from Freiberg, Germany and from Langban, Sweden.

Barytocalcite, 52 mm, Alston Moor, UK

Azurite, 48 mm, Touissit, Morocco

Malachite, 63 mm, Mashamba West, Zair

Azurite
$Cu_3(CO_3)_2(OH)_2$

MONOCLINIC 

Properties: C – blue; S – blue; L – vitreous; D – transparent to opaque; DE – 3.8; H – 3.5; CL – perfect; F – conchoidal; M – tabular and prismatic crystals, pulverulent.
Origin and occurrence: Secondary, resulting from the oxidation of Cu sulfides, mainly associated with malachite. Crystals up to 200 mm (7 7/8 in) long come from Tsumeb, Namibia. Crystals from Touissit, Morocco reach up to 70 mm (2¾ in). Tabular crystals up to 50 mm (2 in) in size found in Chessy near Lyon, France. Famous crystals up to 50 mm (2 in) across occurred in the Copper Queen mine in Bisbee, Arizona, USA. Fine azurite concretions with crystals on the surface come from La Sal, Utah, USA. Crystal rosettes, reaching up to 130 mm (5 2/16 in) in size were found in the Yang Chweng Mine, Guang Dong, China.
Application: Cu ore.

Malachite
$Cu_2(CO_3)(OH)_2$

MONOCLINIC ● ● ● ●

Properties: C – green; S – light green; L – vitreous, dull, earthy; D – opaque; DE – 4.1; H – 3.5-4; CL – perfect; F – conchoidal to uneven; M – acicular and prismatic crystals, botryoidal aggregates, stalactites, massive.
Origin and occurrence: The most common supergene mineral of Cu, associated with azurite, cuprite and other minerals. Crystals up to 30 mm (1 3/16 in) in size occur in Kambove, Zaire. Crystals up to 20 mm (25/32 in) also found in Rudabanya, Hungary. Pseudomorphs after azurite crystals up to 100 mm (4 in) in size known from Tsumeb; pseudo-morphs after cuprite crystals up to 50 mm in size come from Onganja, Namibia. The blocks of banded malachite weighing up to 250 tons, occurred in Mednorudnjansk, Ural mountains, Russia. Similar material in huge quantities come from many deposits in Shaba

Malachite, 70 mm, Shaba, Zair

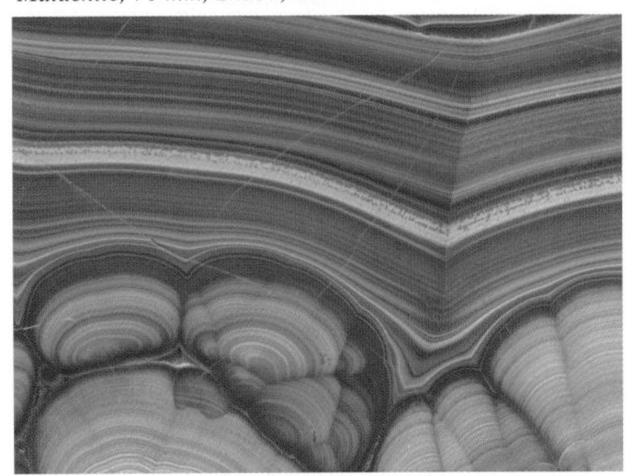

Rosasite, 270 mm, Gleeson, U.S.A.

province, Zaire, where stalactites up to 500 mm (20 in) long were also found.
Application: Cu ore, production of decorative objects and jewelry.

Rosasite

$(Cu,Zn)_2(CO_3)(OH)_2$

MONOCLINIC ● ●

Properties: C – green, blue; S – greenish; D – opaque; DE – 4.0-4.2; H – 4.5; CL – good; M – small acicular crystals, fibrous and botryoidal crusts.
Origin and occurrence: Secondary, forming in the oxidation zone of Cu and Zn deposits, together with other secondary minerals of Cu. Spherical aggregates of acicular crystals up to 10 mm (3/8 in) found in Mina Ojuela, Mapimi, Durango, Mexico. Similar finds were made in Bisbee, Arizona, USA.

Hydrozincite, 65 mm, Tiger, U.S.A.

Hydrozincite

$Zn_5(CO_3)_2(OH)_6$

MONOCLINIC ● ● ●

Properties: C – white, yellowish; S – white; L – pearly to dull; D – opaque; DE – 4; H – 2-2.5; CL – perfect; F – conchoidal; M – tabular crystals, massive, earthy; LU – locally blue.
Origin and occurrence: Secondary, resulting from the oxidation of sphalerite, together with cerussite, smithsonite and hemimorphite. Small crystals are known from Mapimi, Durango, Mexico. Spherical aggregates, several mm in diameter, were found in Sterling Hill, New Jersey, USA. Stalactites and thick crusts occurred in Long-Kieng, Burma. Crusts and stalactites were also described from Bleiberg, Austria; from Mezica, Slovenia and from Raibl, Italy.

Aurichalcite

$(Zn,Cu)_5(CO_3)_2(OH)_6$

ORTHORHOMBIC ● ● ●

Properties: C – light green, blue-green, blue; S – blue-green; L – silky to pearly; D – translucent; DE – 4; H – 1-2; CL – perfect; M – acicular crystals, crusts.
Origin and occurrence: Secondary in the oxidation zone of Cu and Zn deposits in the arid climate, associated with linarite and other minerals. Large prismatic crystals come from Mina Ojuela, Mapimi, Durango, Mexico. Nice rosettes of acicular crystals are known from Bisbee and from the 79 mine, Banner district, Arizona, USA. It also occurred in Monteponi, Sardinia, Italy.

Aurichalcite, 80 mm, Arizona, U.S.A.

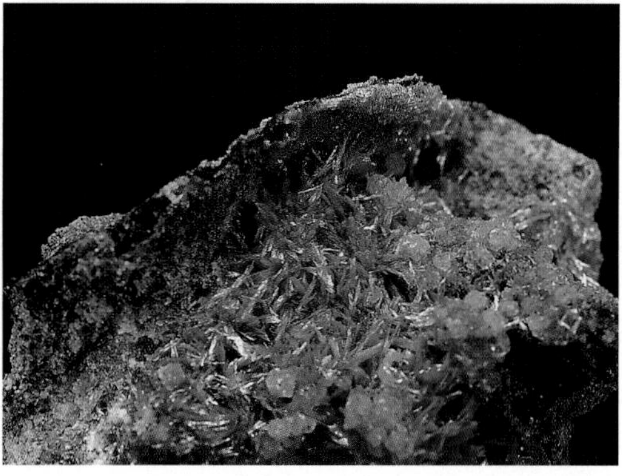

Dawsonite, 2 mm aggregate, Récsk, Hungary

Dawsonite

$NaAl(CO_3)(OH)_2$

ORTHORHOMBIC ● ● ●

Properties: C – colorless to white; S – white; L – vitreous; D – transparent; DE – 2.4; H – 3; CL – perfect; M – acicular to blade-like crystals, radial aggregates.

Origin and occurrence: Hydrothermal in low-temperature deposits, associated with calcite, dolomite and other minerals. Fine acicular crystals up to 35 mm ($1^{3}/_{8}$ in) long come from Mont St.-Hilaire, Quebec, Canada. Radial aggregates along the rock cracks were found in Dubník and Zlatá Bana, Slovakia.

Bastnäsite-(Ce)

$(Ce,La)(CO_3)F$

HEXAGONAL ● ● ●

Properties: C – yellow to brown; S – yellow-brown; L – vitreous to greasy; D – translucent; DE – 4.8-5.2;

Parisite-(Ce), 71 mm, Boyacá, Colombia

Bastnäsite-(Ce), 6 mm x, Ariége, France

H – 4-4.5; CL – good; F – uneven; M – tabular crystals, granular, massive.
Origin and occurrence: Magmatic in pegmatites, also metamorphic, together with allanite-(Ce) and other rare earth elements minerals. Crystals up to 200 x 150 mm (7⁷/₈ x 6 in) in size occur in Andekatany, Madagascar. Crystals up to 100 mm (4 in) long are known from Karonge, Burundi. Transparent crystals up to 25 mm (1 in) long were found recently in the Trimouns quarry, France. It forms important local deposits, as those of Mountain Pass, California, USA.*Application:* ore of rare earth elements.

Parisite-(Ce)

$CaCe_2(CO_3)_3F_2$

TRIGONAL ● ● ●

Properties: C – brown, yellow-brown, gray-yellow; S – brownish; L – vitreous to resinous; D – transparent to translucent; DE – 4.4; H – 4.5; CL – perfect; F – conchoidal to splintery; M – dipyramidal striated crystals.
Origin and occurrence: Magmatic in pegmatites, also hydrothermal and rarely metamorphic, together with bastnäsite-(Ce) and other minerals of rare earth elements. Crystals several cm long come from Quincy, Massachusetts, USA. Crystals up to 80 mm (3¹/₈ in) long occurred in pegmatites near Hundholmen, Norway. Crystals up to 23 cm (9²/₁₆ in) long found in the Snowbird mine, Montana, USA. Transparent crystals up to 15 mm (1⁹/₃₂ in) were lately reported in the Trimouns quarry, France. Very unusual association have been described from Muzo, Columbia, where crystals up to 50 mm occur together with emeralds.

Phosgenite

$Pb_2(CO_3)Cl_2$

TETRAGONAL ● ●

Properties: C – colorless, white, yellow-white, gray, brown; S – white; L – adamantine; D – transparent to translucent; DE – 6.1; H – 2-3; CL – good; F – conchoidal; M – short to long prismatic and tabular crystals, granular, massive; LU – sometimes yellow.
Origin and occurrence: Secondary, resulting from the oxidation of galena, associated with cerussite and other secondary Pb minerals. Crystals up to 150 x 100 mm (6 x 4 in) across known from Monteponi, Sardinia, Italy. Crystals from Tsumeb, Namibia reached up to 100 mm. Crystals up to 30 mm (1³/₁₆ in) in size found in Matlock, Derby, UK. Crystals up to 35 mm (1³/₈ in) long occurred in the Mammoth mine, Tiger, Arizona, USA.

Phosgenite, 28 mm, Monteponi, Italy

Bismutite, 3 mm xx, Brdárka, Slovakia

Gaylussite, 30 mm, Natron Lake, Tanzania

Bismutite

$Bi_2O_2(CO_3)$

TETRAGONAL ● ● ●

Properties: C – yellow, brown, gray, blue, black; S – white; L – vitreous, pearly to dull; D – translucent to opaque; DE – 6.1-7.7; H – 3.5; CL – good; M – spherical, radial aggregates, massive, pulverulent.
Origin and occurrence: Secondary, originated from the oxidation of Bi minerals, associated with bismuthinite and other minerals. It is common from pegmatites in Madagascar (Ampangabé) and in Mozambique. Large pebbles found in Kivu province, Zaire. It also occurred in Tasna, Bolivia.

Weloganite, 30 mm xx, Francon Quarry, Canada

Gaylussite

$Na_2Ca(CO_3)_2 . 5\ H_2O$

MONOCLINIC ● ● ●

Properties: C – colorless, white, gray, yellowish; S – colorless; L – vitreous; D – transparent to translucent; DE – 2; H – 2.5-3; CL – perfect; F – conchoidal; M – lenticular to prismatic crystals.
Origin and occurrence: Sedimentary, typical a constituent of salt sediments. Crystals up to 80 mm (3 1/8 in) long come from Searles Lake, California, USA. It also occurs in Borax Lake and Mono Lake, California, USA. Large crystals are known from Amboseli Lake, Kenya.

Weloganite

$Na_2Sr_3Zr(CO_3)_6 . 3\ H_2O$

TRICLINIC ●

Properties: C – white, yellow; S – white; L – vitreous; D – transparent to opaque; DE – 3.2; H – 3.5; CL – perfect; F – conchoidal; M – pseudohexagonal striated crystals.
Origin and occurrence: Hydrothermal in the alkaline rocks, together with zircon, dresserite and other minerals. Crystals up to 50 mm (2 in) come from cavities in the Francon quarry, Montreal; crystals up to 30 mm (1 3/16 in) are known from St. Michel, Quebec, Canada.

Artinite, 10 mm, Fethiya, Turkey

Hydrotalcite, 1 mm xx, Dunabogdány, Hungary

Artinite

$Mg_2(CO_3)(OH)_2 . 3 H_2O$

MONOCLINIC ● ● ●

Properties: C – white; S – white; L – silky; D – transparent; DE – 2; H – 2.5; CL – perfect; M – spherical, radial aggregates, veinlets, crusts.
Origin and occurrence: Hydrothermal, originating at low temperatures in serpentinites, associated with magnesite, aragonite and other minerals. Radial aggregates up to 20 mm (25/32 in) in size occur in the Gem mine, San Benito County, California, USA. Needles up to 20 mm (25/32 in) long come from Val Malenco, Italy. Clusters of acicular crystals were found on Staten Island, New York, USA.

Zaratite, 70 mm, Lawcaster Co., U.S.A.

Zaratite

$Ni_3(CO_3)(OH)_4 . 4 H_2O$

CUBIC ● ● ●

Properties: C – emerald-green; S – light green; L – vitreous to greasy; D – transparent to translucent; DE – 2.6-2.7; H – 3.5; CL – none; F – conchoidal; M – crystalline crusts, stalactites, coatings.
Origin and occurrence: Secondary, originating from the oxidation of Ni minerals, associated with millerite, brucite and other minerals. It occurs as a product of the oxidation of Ni minerals in ultrabasic rocks in Kraubat and Stubachtal, Austria. Large green coatings found in Heazlewood, Tasmania, Australia. It covers millerite needles in the vicinity of Kladno, Czech Republic.

Hydrotalcite

$Mg_6Al_2(CO_3)(OH)_{16} . 4 H_2O$

TRIGONAL ● ●

Properties: C – white; S – white; L – pearly to waxy; D – transparent; DE – 2.1; H – 2; CL – perfect; M – fibrous and layered aggregates, massive.
Origin and occurrence: : Hydrothermal in ultrabasic and metamorphic rocks. It occurred with serpentine in Nordmark, Norway and also found at Franklin and Sterling Hill, New Jersey, USA.

Stichtite, 35 mm, Rouchovany, Czech Republic

Stichtite

$Mg_6Cr_2(CO_3)(OH)_{16} . 4\ H_2O$

TRIGONAL ● ●

Properties: C – pink to purple; S – white to light purple; L – pearly, waxy to greasy; D – translucent; DE – 2.2; H – 1.5-2; CL – perfect; M – lamellar and fibrous aggregates, massive.
Origin and occurrence: It occurs in serpentinites as scaly aggregates and veinlets in Bou Azzer, Morocco; in Dundas, Tasmania, Australia and in Barberton, South Africa.

Alumohydrocalcite, 70 mm, Ladomírovo, Slovakia

Alumohydrocalcite

$CaAl_2(CO_3)_2(OH)_4 . 3\ H_2O$

TRICLINIC ● ●

Properties: C – white, gray; S – white; L – earthy; D – opaque; DE – 2.2; H – 2.5; CL – perfect; M – chalky aggregates, consisting of acicular crystals.
Origin and occurrence: Secondary, associated with alophane and other minerals. It was described from the Khakasy deposit, Siberia, Russia, where it originates from the oxidation of alophane. Nice white radial aggregates occurred along the cracks in shales in L'adomírovo, Slovakia.

Ancylite-(Ce)

$Sr_3(Ce,La)_4(CO_3)_7(OH)_4 . 3\ H_2O$

ORTHORHOMBIC ● ●

Properties: C – colorless, yellow, yellow- brown, light purple, brown; S – white; L – vitreous to greasy; D – transparent to opaque; DE – 4; H – 4-4.5; CL – none; F – splintery; M – short to long prismatic and pseudo-octahedral crystals.
Origin and occurrence: Hydrothermal in alkaline rocks, associated with nepheline and other minerals. Crystals up to 6 mm (¼ in) long are known from

Ancylite-(Ce), 6 mm xx, Mont St.-Hilaire, Canada

Dresserite, 100 mm, Francon Quarry, Canada

Mont St.-Hilaire, Quebec, Canada. It also occurs in Narssarssuk, Greenland and in Mount Kukisvumchorr, Khibiny massif, Kola Peninsula, Russia.

Dresserite

$BaAl_2(CO_3)_2(OH)_4 . H_2O$

ORTHORHOMBIC ●

Properties: C – white; S – white; L – vitreous to silky; D – transparent; DE – 3; H – 2.5-3; M – radial and spherical aggregates, consisting of acicular crystals. *Origin and occurrence:* Hydrothermal in alkaline rocks, together with weloganite. Clusters of acicular crystals up to 4 mm (5/32 in) long occur in the Francon quarry, Montreal, Quebec, Canada.

Andersonite

$Na_2Ca(UO_2)(CO_3)_3 . 6 H_2O$

TRIGONAL ● ●

Properties: C – yellow-green; S – yellow; L – vitreous to pearly; D – transparent to translucent; DE – 2.9; H – 2.5; M – rhombohedral crystals, crusts, veinlets; LU – intensive yellow-green; R – radioactive.

Origin and occurrence: Secondary, originated from the oxidation of primary U ores, associated with other U minerals. Crystals up to 10 mm (3/8 in) in size found in the Ambrosia Lakes district, New Mexico, USA. It occurs as coatings on the walls of the tunnels in the Hillside mine near Bagdad, Arizona, USA. Crystals up to 10 mm (3/8 in) across and thick crusts come from the Atomic King No.2 mine near Moab, Utah, USA.

Andersonite, 80 mm, Repete Mine, Utah, U.S.A.

6. Borates

Ludwigite

$Mg_2Fe^{3+}BO_5$

ORTHORHOMBIC ● ● ●

Properties: C – dark green, black-green, black; S – blue-green; L B silky to dull; D – opaque; DE – 3.9; H – 5; CL – perfect; F B uneven; M – prismatic crystals, fibrous aggregates, granular, massive.
Origin and occurrence: Metamorphic in skarns and dolomitic marbles, locally associated with magnetite and other borates. Rich aggregates are known from Ocna de Fier, Romania; Kamineichi, Japan; and the Hol Kol mine, Suan, North Korea.
Application: chemical industry, B ore.

Gaudefroyite

$Ca_4Mn^{3+}{}_3(BO_3)_3CO_3(O,OH)_3$

HEXAGONAL ●

Properties: C – black; S – black; L – adamantine to dull; D – opaque; DE – 3.4; H – 6; CL – good; F – uneven; M – prismatic crystals, fibrous and acicular aggregates.
Origin and occurrence: Hydrothermal in calcite veins. Prismatic crystals up to 50 mm (2 in) long were found on the mine dumps near Tachgagalt, Morocco and in the N'Chwaning No. 2 mine and the Wessels mine, Kuruman, South Africa.

Gaudefroyite, 40 mm, Kuruman, South Africa
Ludwigite, 120 mm, Ocna de Fier, Romania

Inderite, 230 mm, Boron, U.S.A.

Inderite

$MgB_3O_3(OH)_5 \cdot 5\,H_2O$

MONOCLINIC ● ● ●

Properties: C – colorless, white to pink in aggregates; S – white; L – vitreous; D – transparent to translucent; DE – 1.8; H – 2.5; CL – good; F – conchoidal to uneven; M – prismatic crystals, acicular and fibrous aggregates, nodules, massive.
Origin and occurrence: Sedimentary in boron deposits, commonly associated with colemanite and other borates. Prismatic crystals up to 100 mm (4 in) long occur in Boron, California, USA. It is also known from Inder, Kazakhstan.
Application: chemical industry, B ore.

Inyoite, 90 mm, Turkey

Borax, 60 mm x, Searles Lake, U.S.A.

Inyoite

$CaB_3O_3(OH)_5 . 4 H_2O$

MONOCLINIC ● ● ●

Properties: C – colorless, white; S – white; L – vitreous; D – transparent to translucent; DE B 1.9; H – 2; CL – good; F – uneven; M – short prismatic to tabular crystals, columnar aggregates, massive.

Origin and occurrence: Sedimentary in boron deposits, commonly associated with colemanite and other borates. Clear tabular crystals up to 100 mm (4 in) long found in Kirka and Emet, Turkey. It also comes from Inder, Kazakhstan and the Corkscrew mine, California, USA.

Application: chemical industry, B ore.

Ulexite, 20 mm aggregate, Death Valley, California, U.S.A.

Colemanite, 276 mm, Death Valley, California, U.S.A.

Borax

$Na_2B_4O_5(OH)_4 . 8\ H_2O$

MONOCLINIC ● ● ●

Properties: C – white, colorless, yellowish, gray, greenish; S – white; L B vitreous to dull; D – transparent, translucent to opaque; DE – 1.7; H – 2-2.5; CL – perfect; F – conchoidal; M – short prismatic to tabular crystals, columnar to earthy aggregates, crusts, coatings, granular, massive; R – soluble in water.
Origin and occurrence: Sedimentary in boron deposits, associated with other borates and halite. Prismatic crystals up to 150 mm long are known from Borax Lake, also Searles Lake and Boron, California, USA. It also occurs in Kirka, Turkey.
Application: chemical industry, B ore.

Ulexite

$NaCaB_5O_6(OH)_6 . 5\ H_2O$

TRICLINIC ● ● ●

Properties: C – white, colorless, light gray; S – white; L B vitreous to silky; D – transparent to translucent; DE – 2.0; H – 2.5; CL – perfect; F B uneven; M – elongated prismatic to acicular crystals, fibrous aggregates, crusts, nodules, granular, massive.
Origin and occurrence: Sedimentary in boron deposits, associated with other borates. Slabs up to 100 mm (4 in) thick consisting of fibrous aggregates occur in Boron, California; also found in Esmeralda, Nevada, USA; Emet, Turkey; and Inder, Kazakhstan.
Application: chemical industry, B ore.

Colemanite

$Ca_2B_6O_{11} . 5\ H_2O$

MONOCLINIC ● ● ●

Properties: C – colorless, white, yellowish, light gray; S – white; L – vitreous; D – transparent to translucent; DE – 2.4; H – 4.5; CL – perfect; F B uneven; M – short prismatic and isometric crystals, granular, massive.
Origin and occurrence: Sedimentary in boron deposits, associated with other borates. Perfect crystals up to 200 mm (7$\frac{7}{8}$ in), come from Emet and Kirka, Turkey. Crystals up to 70 mm (2$\frac{3}{4}$ in) long occur in Gower Gulch, Inyo, California; also known from Esmeralda, Nevada, USA and Inder, Kazakhstan.
Application: chemical industry, B ore.

Kernite, 100 mm, Boron, U.S.A.

Hambergite, 64 mm, Pamir, Tadzhikistan

Kernite

$Na_2B_4O_6(OH)_2 . 3 H_2O$

MONOCLINIC ● ● ●

Properties: C – colorless, white, light gray; S – white; L – vitreous to dull, silky; D – transparent, translucent to opaque; DE – 1.9; H – 2.5-3; CL – perfect; F – uneven; M – isometric crystals, fibrous aggregates, granular, massive; R – soluble in water.

Origin and occurrence: Sedimentary in boron deposits, associated with borax and other borates. Platy aggregates and crystals up to 2.5 x 1 m (8 ft x 39 3/8 in) found in Boron, California, USA; also known from the Tincalayu mine, Salta province, Argentina.

Rhodizite, 23 mm, Antsirabé, Madagascar

Hambergite

$Be_2BO_3(OH,F)$

ORTHORHOMBIC ● ●

Properties: C – colorless, white, light gray; S – white; L – vitreous; D – transparent to translucent; DE – 2.4; H – 7.5; CL B perfect; F B uneven; M – tabular to prismatic crystals, granular.

Origin and occurrence: Magmatic in granitic and rarely also in alkaline pegmatites; hydrothermal in cavities within the pegmatites, associated with tourmaline, danburite and beryl. Tabular crystals up to 200 mm (7 7/8 in) from the Little Three mine, Ramona, California, USA. Crystals up to 110 mm (4 5/16 in) also known from several localities in Madagascar, e.g. Imalo and Anjanabonoina; also found in Hyakule, Nepal and Ctidružice, Czech Republic .

Rhodizite

$(K,Cs)Al_4Be_4(B,Be)_{12}O_{28}$

CUBIC ● ●

Properties: C – colorless, white, yellow, light gray; S – white; L – vitreous to adamantine; D – transparent to translucent; DE – 3.5; H – 8.5; CL – imperfect; F – conchoidal to uneven; M – isometric crystals, granular.

Origin and occurrence: Magmatic in granitic pegmatites, associated with tourmaline, danburite and beryl. Cubic to tetrahedral crystals, reaching up to 30 mm (1 3/16 in) occur in several localities in

Madagascar, e.g. Sahatany and Antandrokomby; also known from the Animikie Red Ace pegmatite, Florence, Wisconsin, USA.

Boracite

$Mg_3B_7O_{13}Cl$

ORTHORHOMBIC ● ● ●

Properties: C – colorless, white, yellowish, light gray, light to dark green; S – white; L – vitreous; D – transparent to translucent; DE – 3.0; H – 7-7.5; CL – none; F – conchoidal to uneven; M – isometric crystals, fibrous aggregates, granular, massive.

Origin and occurrence: Sedimentary in evaporitic deposits, together with halite, gypsum and anhydrite; metamorphic in metamorphosed evaporates. Crystals up to 15 mm (19/32 in) come from Alto Chapare, Cochabamba, Bolivia; crystals about 5 mm (3/16 in) in size, occur in Stassfurt, Hanover and Kahlberg, Germany. Crystals are also known from Choctaw, Louisiana, USA and the Boulby mine, North Yorkshire, England, UK.

Boracite, 12 mm x, Alto Chapare, Bolivia

7. Sulfates

Anglesite

$PbSO_4$

ORTHORHOMBIC ● ● ●

Properties: C – colorless, white, yellowish, gray, greenish; S – white; L – adamantine to greasy; D – transparent to translucent; DE – 6.4; H – 2.5 – 3; CL – dobrá; F – conchoidal; M – thick tabular crystals, massive, LU – sometimes yellowish.

Origin and occurrence: Secondary, as a result of the galena oxidation, together with cerussite and other minerals. Beautiful yellowish crystals up to 100 mm (4 in) in size come from Touissit, Morocco. Crystals from Tsumeb, Namibia, reached up to 40 mm (1 9/16 in). Large crystals are also reported from Phoenixville, Pennsylvania, USA. Crystals up to 20 mm (25/32 in) occur in cavities in weathered galena in Sardinia, Italy. Crystals, reaching up to 40 mm (1 9/16 in) also found in Mežica, Slovenia. Prismatic crystals up to 75 mm (3 in) long known from the Bunker Hill mine, Idaho, USA.

Barite, 17 mm xx, Pöhla, Germany
Anhydrite, 55 mm, Simplon Tunnel, Switzerland

Anglesite, 81 mm, Touissit, Morocco

Anhydrite

$CaSO_4$

ORTHORHOMBIC ● ● ● ●

Properties: C – colorless, white, bluish, purplish, red, brown; S – white; L – pearly to vitreous; PS – transparent to translucent; DE – 2.8-3; H – 3-3.5; CL – good; F – splintery to uneven; M – isometric and prismatic crystals, granular, massive.

Origin and occurrence: Mostly sedimentary, as a result of the evaporation of sea water, associated with gypsum, calcite and other minerals; rare in pegmatites and hydrothermal. It is very common in form of massive aggregates in salt deposits in Stassfurt and Wathlingen, Germany, where small crystals also occur. Folded layers in clays known from Wieliczka, Poland. Purplish crystals up to 20 mm (25/32 in) found in cracks in metamorphic rocks near Simplon and St. Gotthard, Switzerland. Fine druses of bluish crystals up to 200 mm (7 7/8 in) long found in Naica, Chihuahua, Mexico.

Celestine, 270 mm, Madagascar

Celestine
$SrSO_4$

ORTHORHOMBIC ● ● ●

Properties: C – colorless, white, yellowish, blue; S – white; L – vitreous; D – transparent to translucent; DE – 4; H – 3-3.5; CL – perfect; F – uneven; M – prismatic and tabular crystals, columnar aggregates, concretions.
Origin and occurrence: Rare hydrothermal, mainly sedimentary, together with halite, anhydrite and gypsum. Beautiful blue crystals several cm long occur in cavities of concretions in the Sakoany mine, Madagascar. Prismatic crystals up to 100 mm (4 in) long in sulfur deposits in Poland (Tarnobrzeg) and Italy (Caltanissetta). Fine blue tabular crystals associated with aragonite crystals found in Spania Dolina, Slovakia. Fine crystals are known from marls in Tunisia and Libya.

Barite
$BaSO_4$

ORTHORHOMBIC ● ● ● ●

Properties: C – white, yellow, blue, red, brown, black; S – white; L – vitreous; D – transparent to translucent; DE – 4.3-4.7; H – 3.5; CL – perfect; F – conchoidal to splintery; M – tabular to prismatic crystals, massive; LU – sometimes blue.
Origin and occurrence: Hydrothermal, it originates under medium and low temperatures, also sedimentary, together with fluorite, calcite, cinnabar and other minerals. Beautiful barite crystals up to 50 mm (2 in) found in hydrothermal veins in Příbram, Czech Republic. Blue tabular crystals up to 100 mm (4 in) from Dědova hora, Czech Republic. Beautiful tabular crystals up to 100 mm (4 in) known from Banská Štiavnica, Slovakia; Cavnic and Baia Sprie, Romania. Very fine crystals of up to 200 mm ($7^{7}/_{8}$ in) occurred in Alston Moor, Frizington and Mowbray, Cumbria, UK. Beautiful druses of honey-brown crystals up to 70 mm (2¾ in) long found in Pöhla, Germany. Fine druses of prismatic crystals known from Elk Creek, South Dakota, USA; so called

Antlerite, 60 mm, Chuquicamata, Chile

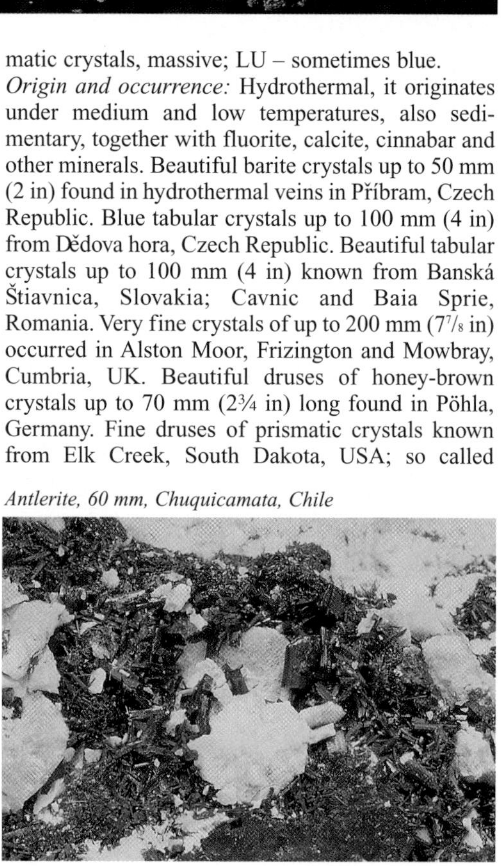

Barite, 100 mm xx, Stoneham Co., U.S.A.

desert roses (crystals with inclusion of sand grains) originating in desert climate occur in the vicinity of Norman, Oklahoma, USA. Shiny crystals come from Freiberg and Halsbrücke, Germany.
Application: an ingredient of drilling fluids, chemical, glass, paper and rubber industries.

Antlerite

$Cu_5(SO_4)_2(OH)_6 \cdot 3\ H_2O$

ORTHORHOMBIC ● ● ●

Properties: C – green; S – light green; L – vitreous; D – translucent; DE – 3.9; H – 3.5; CL – perfect; F – ćonchoidal to uneven; M – short prismatic and acicular crystals, crusts and earthy aggregates.
Origin and occurrence: Secondary, resulting from weathering of Cu ores, associated with other secondary Cu minerals. It is a principal mineral of the oxidation zone in Chuquicamata, Chile, where it forms crystals up to 5 mm ($^3/_{16}$ in). Acicular crystals, up to 20 mm ($^{25}/_{32}$ in) long, occurred in Bisbee, Arizona, USA. Crystalline crusts are known from Špania Dolina and Piesky, Slovakia.

Brochantite, 14 mm aggregate, Bingham, U.S.A.

Brochantite

$Cu_4(SO_4)(OH)_6$

MONOCLINIC ● ● ●

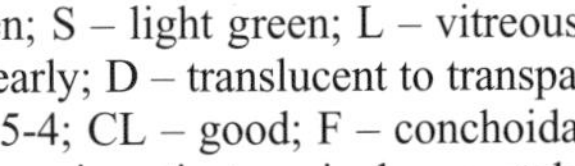

Properties: C – green; S – light green; L – vitreous, in cleavage planes pearly; D – translucent to transparent; DE – 4; H – 3.5-4; CL – good; F – conchoidal to uneven; M – long prismatic to acicular crystals, granular.
Origin and occurrence: Secondary, as a result of the oxidation of Cu ores usually in the arid climate, together with other secondary Cu minerals. Crystals up to 70 mm (2¾ in) long found in Bisbee, Arizona, USA, together with pseudo-morphs of malachite after brochantite. Crystals are also known from Tsumeb, Namibia. Prismatic crystals up to 50 mm (2 in) long from Cerro Verde, Peru. Emerald-green crusts occur in L'ubietová, Slovakia.

Linarite, 23 mm xx, Graham Co., U.S.A.

Linarite

$PbCu(SO_4)(OH)_2$

MONOCLINIC ● ●

Properties: C – azure-blue; S – light blue; L – vitreous to adamantine; D – translucent; DE – 5.3-5.5; H – 2.5; CL – good; F – conchoidal; M – tabular and prismatic crystals, coatings.

Origin and occurrence: Secondary, as a result of the oxidation of Cu and Pb sulfides at low pH. The largest crystals, reaching up to 80 mm (3¹/₈ in), come from the Mammoth mine, Tiger, Arizona, USA. Crystals from the Grand Reef mine in Arizona up to 50 mm (2 in) long. Very fine crystals, up to 30 mm (1³/₁₆ in) were found in Keswick, Cumbria, England, UK. Large crystals were also described from Tsumeb, Namibia and from Broken Hill, New South Wales, Australia.

Alunite

$KAl_3(SO_4)_2(OH)_6$

TRIGONAL ● ● ●

Properties: C – white, yellowish, gray; S – white; L – vitreous to pearly; D – transparent to opaque; DE – 2.8; H – 3.5-4; CL – perfect; F – conchoidal; M – rhombohedral crystals, porous aggregates, granular.

Origin and occurrence: Secondary, as a result of reactions of sulfuric acid with Al-rich rocks, associated with gypsum. Small crystals are known in Berehovo, Ukraine. Massive aggregates occur together with turquoise in Cu deposits in Arizona, USA. (New Cornelia mine and others). Huge alunite deposits, like Zaglik, Azerbaijan, are mined for Al.

Application: Al ore.

Alunite, 1 mm xx, Nagyegyháza, Hungary

Natrojarosite, 50 mm, Laurion, Greece

Natrojarosite

$NaFe^{3+}_3(SO_4)_2(OH)_6$

TRIGONAL ● ●

Properties: C – yellow, brown; S – light yellow; L – vitreous; D – transparent to translucent; DE – 3.2; H – 3; CL – good; F – conchoidal; M – tabular to rhombohedral crystals, earthy aggregates..

Origin and occurrence: Secondary, as a result of weathering Fe sulfides, associated with fibroferrite, alunite and other minerals. It occurs in Modum, Norway; Soda Springs Valley, Nevada, USA; Chuquicamata, Chile and elsewhere.

Jarosite

$KFe^{3+}_3(SO_4)_2(OH)_6$

TRIGONAL ● ● ●

Properties: C – yellow; S – light yellow; L – adamantine to dull; D – translucent; DE – 2.9-3.3; H – 2.5-3.5; CL – good; F – conchoidal to uneven; M – rhombohedral to tabular crystals, granular, massive.

Origin and occurrence: : Secondary, as a result of weathering Fe sulfides, usually associated with natrojarosite and other sulfates. The world's best specimens come from Peña Blanca Uranium mine, near Aldama, Chihuahua, Mexico, where crystals, up to 20 mm (25/32 in) were found. Crystals, up to 10 mm (3/8 in), were found in Tombstone, Arizona, USA. Crystals, together with pseudo-morphs of jarosite after alunite, occurred in Chuquicamata, Chile. Tabular crystals also were described from Horní Slavkov, Czech Republic.

Beudantite

$PbFe^{3+}_3(AsO_4)(SO_4)(OH)_6$

TRIGONAL ● ●

Properties: C – dark green, brown, red-brown; S – greenish, gray-yellow; L – vitreous to resinous; D – transparent to translucent; DE – 4; H – 3.5-4.5; CL – good; M – rhombohedral crystals, crusts.

Origin and occurrence: Secondary, occurring in the oxidation zone of Pb deposits, together with scorodite and other minerals. Crystals, several mm in size, found in Tsumeb, Namibia, are the best in the world. Small crystals are known from Bisbee, Arizona, USA; Ashburton Downs, Western Australia; and Kamaréza near Laurion, Greece.

Jarosite, 2 mm xx, Gyöngyösoroszi, Hungary

Beudantite, 70 mm, Moldava, Czech Republic

Hanksite

$Na_{22}K(SO_4)_9(CO_3)_2Cl$

HEXAGONAL ● ●

Properties: : C – colorless, yellowish; S – colorless; L – vitreous to dull; D – transparent to translucent; DE – 2.6; H – 3.5; CL – good; F – uneven; M – tabular to short prismatic crystals.
Origin and occurrence: Sedimentary in salt lake sediments, together with halite, borax and other minerals. Crystals up to 200 mm (7⁷/₈ in) found in Searles Lake, California, USA. It is also known from borax deposits in the Death Valley, California, USA.

Caledonite

$Cu_2Pb_5(SO_4)_3(CO_3)(OH)_6$

ORTHORHOMBIC ● ●

Properties: C – green to blue-green; S – light green; L – resinous; D – transparent to translucent; DE – 5.6; H – 2.5-3; CL – perfect; F – uneven; M – prismatic crystals, radial aggregates.

Origin and occurrence: Secondary, occurring in the oxidation zone of Cu and Pb deposits, associated with linarite and other minerals. Crystals up to 20 mm ($^{25}/_{32}$ in) come from Leadhills, Scotland, UK. Crystals, up to 15 mm ($^{19}/_{32}$ in), were found in the Mammoth mine, Tiger, Arizona, USA. Rich druses are known from Anarak, Iran. Crystals reaching up to 20 mm ($^{25}/_{32}$ in) occurred in the Blue Bell mine, California, USA.

Leadhillite

$Pb_4(SO_4)(CO_3)_2(OH)_2$

MONOCLINIC ● ●

Properties: C – colorless, yellowish, gray; S – colorless; L – resinous to adamantine; D – transparent to translucent; DE – 6.5; H – 2.5-3; CL – perfect; F – conchoidal; M – pseudo-hexagonal tabular crystals.
Origin and occurrence: Secondary mineral from the oxidation zone of Pb deposits, associated with cerussite, anglesite, linarite and other minerals. The largest crystals, measuring up to 15 cm, come from Tsumeb, Namibia. Crystals up to 25 mm (1 in) found in the Mammoth mine, Tiger, Arizona, USA. Crystals in Leadhills, Scotland, UK were of similar size.

Chalcantite

$CuSO_4 . 5 H_2O$

TRICLINIC ● ● ●

Properties: C – deep blue; S – white; L – vitreous to resinous; D – transparent to translucent; DE – 2.3; H – 2.5; CL – imperfect; F – conchoidal; M – short prismatic to tabular crystals, stalactites, films; R – soluble in water.

Hanksite, 50 mm, Searles Lake, U.S.A.

Caledonite, 5 mm x, Graham Co., U.S.A.

Leadhillite, 33 mm, Leadhill, UK

Chalcanthite, 36 mm, Pima Co., U.S.A.

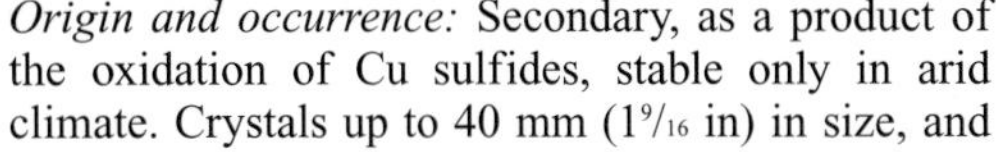

Origin and occurrence: Secondary, as a product of the oxidation of Cu sulfides, stable only in arid climate. Crystals up to 40 mm ($1^{9}/_{16}$ in) in size, and stalactites, up to 1 m ($39^{3}/_{8}$ in) long are known from Bisbee, Arizona, USA. Crystals were also found in the oxidation zone in Chuquicamata, Chile. Beautiful stalactites come from Rio Tinto, Spain.
Application: rarely mined as Cu ore.

Melanterite, 80 mm, Harz Mts., Germany

Melanterite

$FeSO_4 . 7 H_2O$

MONOCLINIC ● ● ● ●

Varieties: pisanite (with Cu contents), kirovite (with Mg contents)

Properties: C – light green; S – colorless; L – vitreous; D – translucent; DE – 1.9; H – 2; CL – perfect; F – conchoidal; M – granular crusts, botryoidal and stalactitic aggregates, films; R – soluble in water.
Origin and occurrence: Secondary, as a result of the oxidation of Fe sulfides, together with other sulfates. It is unstable under atmospheric conditions. Crystals up to 20 mm ($^{25}/_{32}$ in) long are known from Bisbee, Arizona and from the Boyd mine, Tennessee, USA. Stalactites, up to 50 cm (20 in) long, occurred in Chvaletice, Czech Republic. It is also common in Rio Tinto, Spain and Banská Štiavnica, Slovakia.

Epsomite
$MgSO_4 \cdot 7\ H_2O$

ORTHORHOMBIC ● ● ● ●

Properties: C – colorless to white; S – colorless; L – vitreous to silky; D – transparent to translucent; DE – 1.7; H – 2; CL – perfect; F – conchoidal; M – small crystals, granular, stalactitic aggregates; R – soluble in water, decomposing under atmospheric conditions.
Origin and occurrence: Secondary, as a result of the oxidation of Fe sulfides, also from crystallization of the salt lake water, associated with halite and other minerals. Crystals up to 1 m (39³/₈ in) long were found on Mount Kruger, Washington; prismatic crystals up to 50 mm (2 in) long also occur in Bisbee, Arizona and needles of a similar size come from the White Caps mine, Nevada, USA. It is also known as a product of the activity of fumaroles in Mount Vesuvius, Italy.

Morenosite
$NiSO_4 \cdot 7\ H_2O$

ORTHORHOMBIC ● ● ●

Properties: C – green, green-white; S – greenish; L – vitreous; D – transparent to translucent; DE – 2; H – 2-2.5; CL – good; F – conchoidal; M – stalactitic crusts and efflorescences; R – soluble in water.
Origin and occurrence: Secondary, as a product of the oxidation of Ni minerals, stable mainly in the arid regions. It is very common in Sudbury, Ontario, Canada; also in Richelsdorf, Germany and Potůčky, Czech Republic.

Coquimbite
$Fe^{3+}_2(SO_4)_3 \cdot 9\ H_2O$

TRIGONAL ● ● ●

Properties: C – purple, yellow, green, colorless; S – colorless; L – vitreous; D – transparent; DE – 2.1; H – 2; CL – imperfect; F – conchoidal to uneven; M – prismatic to tabular crystals, granular, massive; R – soluble in water.
Origin and occurrence: Secondary, associated with other sulfates. Crystals, several cm long, were found in the Dexter No.7 mine, Calf Mesa, Utah, USA. Small prismatic crystals are known from Železník, tabular crystals occurred in Banská Štiavnica, Slovakia. It is very common in Chilean deposits, like Chuquicamata and Tierra Amarilla.

Alunogen
$Al_2(SO_4)_3 \cdot 17\ H_2O$

TRICLINIC ● ● ●

Properties: C – colorless, white, gray-yellow; S – colorless; L – silky; PS – transparent; DE – 1.8; H – 1.5; CL – perfect; M – pseudo-hexagonal crystals, granular; R – soluble in water.
Origin and occurrence: Secondary, as a result of the pyrite oxidation, also a product of sublimation on volcanoes and burning dumps, associated with other sulfates. Large crystals are known from the Dexter No.7 mine, Calf Mesa, Utah, crusts, over 1 m (39³/₈ in) thick were found on Mount Alum, New Mexico,

Epsomite, 10 mm xx, Stassfurt, Germany

Morenosite, 60 mm, Potůčky, Czech Republic

Coquimbite, 110 mm, Cerritos Bajos, Chile

Alunogen, 80 mm, Dubník, Slovakia

USA. Fine aggregates occur on the walls of underground workings in the old opal mines in Dubník, Slovakia.

Halotrichite

$FeAl_2(SO_4)_4 . 22 H_2O$

MONOCLINIC ● ● ●

Properties: C – white, greenish; S – colorless; L – vitreous; PS – transparent to translucent; DE – 1.9-2.1; H – 1.5-2; CL – imperfect; F – conchoidal; M – acicular crystals, fibrous aggregates, efflorescences; R – soluble in water.
Origin and occurrence: Secondary, as a result of the pyrite oxidation, also a product of the activity of hot springs ans solfataras, associated with other sulfates. Common fibrous crusts occur in Dubník, Slovakia. It looks similar in Rio Marina, Elba, Italy and in Chuquicamata, Chile. It is known from solfataras in Pozzuoli, Italy. It is a product of hot springs activity in the Lassen Peak National Park, California, USA.

Halotrichite, 90 mm, Récsk, Hungary

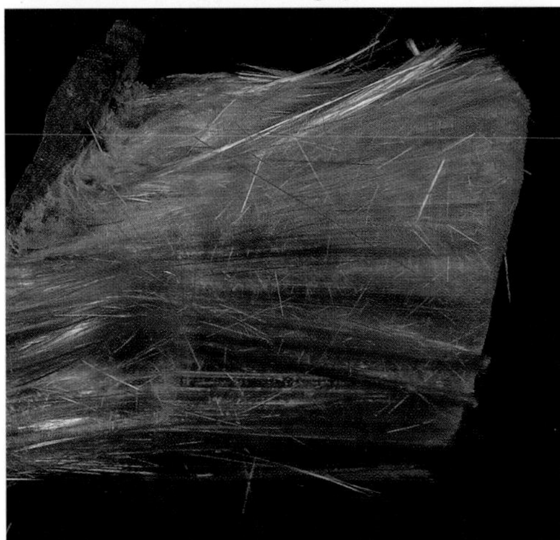

Apjohnite

$MnAl_2(SO_4)_4 . 22 H_2O$

MONOCLINIC ● ● ●

Properties: C – colorless, white, pink, greenish, yellow; S – colorless; L – silky; D – transparent; DE – 1.9; H – 1.5; M – fibrous aggregates, coatings, massive; R – soluble in water.
Origin and occurrence: Secondary, occurring together with other sulfates. Large accumulations are known from Little Pigeon Creek, Alum Cave, Tennessee, USA. It also occurs in Delagona Bay, Mozambique.

Apjohnite, 60 mm, Smolník, Slovakia

Bílinite, 60 mm, Bílina, Czech Republic

Bilinite

$Fe^{2+}Fe^{3+}_2(SO_4)_4 . 22 H_2O$

MONOCLINIC ● ●

Properties: C – white to wellowish; S – white; L – silky; D – opaque; DE – 1.9; H – 2; M – fibrous aggregates; R – soluble in water.
Origin and occurrence: Secondary, as a result of the pyrite oxidation, associated with other sulfates. Fibrous aggregates are known from Svitec near Bílina, Czech Republic. Also discovered in Bisbee, Arizona, USA.

Tschermigite

$(NH_4)Al(SO_4)_2 . 12 H_2O$

CUBIC ● ● ●

Properties: : C – colorless; S – colorless; L – vitreous; D – transparent; DE – 1.6; H – 1.5; F – conchoidal; M – octahedral crystals, fibrous and columnar aggregates; R – soluble in water.
Origin and occurrence: Secondary in the brown coal, also on burning dumps and as a product of solfataras. Skeletal crystals up to 10 mm (3/8 in) in size are known from the brown coal basin near Most and Bílina, Czech Republic. Original locality Cermníky, Czech Republic yielded fibrous crusts, up to 50 mm (2 in) thick. Fine octahedra, up to 10 mm (3/8 in) come from burning dumps near Zastávka near Brno, Czech Republic. Octahedra occur also in Geysers, California, USA. Small crystals on Etna, Sicily, Italy are of volcanic origin.

Tschermigite, 50 mm, Mogyorórosbánya, Hungary

Polyhalite

$KCa_2Mg(SO_4)_4 . 2 H_2O$

TRICLINIC ● ● ●

Properties: C – colorless, brown, red-brown; S – colorless; L – vitreous; D – transparent to translucent; DE – 2.8; H – 3-3.5; CL – perfect; M – acicular to prismatic crystals, columnar, scaly and fibrous aggregates.
Origin and occurrence: Sedimentary, as a constituent of salt deposits, also as a product of volcanic activity. Tabular and prismatic crystals are found rarely together with massive polyhalite aggregates in Stassfurt, Germany. Fibrous aggregates come from Halstatt, Austria. Large deposits of polyhalite are located near Carlsbad, New Mexico, USA. Coatings occur on Mount Vesuvius, Italy.

Görgeyite

$K_2Ca_5(SO_4)_6 . H_2O$

MONOCLINIC ● ●

Properties: C – colorless, yellowish; S – colorless; L – vitreous; D – translucent; DE – 3; H – 3.5;

Polyhalite, 54 mm, Carlsbad, U.S.A.

Gypsum, 62 mm, Carneville, Utah, U.S.A.

CL – good; F – conchoidal; M – tabular crystals.
Origin and occurrence: Sedimentary in salt deposits, associated with halite and other minerals. Fine crystals up to 80 mm ($3^{1}/_{8}$ in) long found near Inder Lake, Kazakhstan; also known from Bad Ischl, Austria.

Gypsum
$CaSO_4 . 2 H_2O$

MONOCLINIC ● ● ● ● ●

Varieties: Maria-glass, alabaster, satin spar, selenite.
Properties: C – colorless, white, gray, yellowish; S – colorless; L – vitreous to pearly; D – transparent to translucent; DE – 2.3; H – 1.5-2; CL – perfect; F – conchoidal; M – typical monoclinic crystals, often twins, fibrous and platy aggregates, granular, massive; LU – crystals with inclusions sometimes bluish to yellowish.

Görgeyite, 80 mm, Inder Lake, Kazakhstan

Origin and occurrence: Rare primary hydrothermal, mostly sedimentary and as a weathering product, associated with anhydrite, halite and other minerals. Crystals up to 50 mm from Cavnic, Romania are probably hydrothermal. Huge crystals up to 1.5 m (5 ft) long found in karst cavities (Cave of Swords) in gossan in the PbBZn deposit Naica, Chihuahua, Mexico. Crystals up to 9 m (29 ft 6 in) long from Santa Eulalia, Chihuahua, Mexico, where they were found combined with interesting aggregates, called 'ram's horns' from their shape. Crystals up to 4 m (13 ft) long occurred in Tarnobrzeg, Poland. Common crystals are also known from Gorguel, Spain. The 'desert roses' or crystal druses, from Sahara desert in Tunisia and Algeria, with sand grain inclusions are mineralogically interesting. Large slabs of the transparent variety, called Maria-glass, found in Friedrichsrode, Germany. Fine-grained variety alabaster occurs e.g. in Italy. Fibrous variety selenite (sometimes also called satin spar) comes from the Sylva river basin, Perm, Russia.
Application: building, chemical and medical industries.

Langite
$Cu_4(SO_4)(OH)_6 . 2 H_2O$

ORTHORHOMBIC ● ● ●

Properties: C – blue, blue-green; S – light blue; L – vitreous to silky; D – translucent; DE – 3.3; H – 2.5-3; CL – perfect; M – small isometric crystals, fine-grained crusts, earthy aggregates.
Origin and occurrence: Secondary, as a product of the oxidation of Cu ores, associated with gypsum and other minerals. Crystals known from St. Just, Cornwall, UK. Crystals occurred in Tsumeb, Namibia. Fine specimens come from Špania Dolina and L'ubietová, Slovakia and from Borovec, Czech Republic. It is also common in Ely, Nevada, USA and in El Cobre, Chile.

Langite, 49 mm, Allihies, Cork, Ireland

Fibroferrite, 70 mm, Dubník, Slovakia

Aluminite, 60 mm, Bicske-Csordakút, Hungary

Fibroferrite

$Fe^{3+}(SO_4)(OH) . 5 H_2O$

ORTHORHOMBIC ● ● ●

Properties: C – yellowish, greenish, gray; S – white; L – silky; D – opaque; DE – 1.9; H – 2.5; CL – perfect; M – small crystals, fibrous and botryoidal crusts; R – soluble in water.

Origin and occurrence: Secondary as a product of the pyrite oxidation, together with other sulfates. Crystals in cavities in melanterite found in the Dexter No. 7 mine, Calf Mesa, Utah, USA. Vein fillings up to 3 m (10 ft) thick come from the Santa Elena mine, La Alcaparrosa, Argentina. Fine fibrous aggregates occur in Dubník, Slovakia. It is common in many localities in Chile (Tierra Amarilla, Chuquicamata).

Aluminite

$Al_2(SO_4)(OH)_4 . 7 H_2O$

MONOCLINIC ● ● ●

Properties: C – white; S – white; L – dull; D – opaque; DE – 1.7-1.8; H – 1-2; CL – none; M – earthy nodules, consisting of microscopic needles.

Origin and occurrence: Secondary, resulting from the reaction of sulfuric acid with Al in Al-rich rocks. Large nodule, 30 cm (12 in) in diameter, known from Newhaven, Sussex, UK. Fine white nodules described from Malá Chuchle, Prague, Czech Republic. It also occurs near Halle, Germany and covers limestone in the vicinity of Joplin, Missouri, USA.

Botryogen

$MgFe^{3+}(SO_4)_2(OH) . 7 H_2O$

MONOCLINIC ● ● ●

Properties: C – orange-red; S – yellow; L – vitreous; D – transparent to translucent; DE – 2.1; H – 2-2.5; CL – perfect; F – conchoidal; M – prismatic striated crystals, racemose and spherical aggregates with radial structure.

Origin and occurrence: Secondary, as a result of the pyrite oxidation in arid regions, together with other sulfates. Crystals up to 35 mm ($1^{5}/_{16}$ in) long from the Libiola mine near Genoa, Italy. It is common in Chuquicamata and Quetena in Chile; and is also known to be from Rammelsberg, Germany.

Botryogen, 3 mm xx, Knoxville, U.S.A.

Devilline, 10 mm xx, Špania Dolina, Slovakia

Copiapite

$(Fe^{2+},Mg)Fe^{3+}{}_4(SO_4)_6(OH)_2 \cdot 20\ H_2O$

TRICLINIC ● ● ●

Properties: C – yellow, green-yellow, orange; S – yellow; L – pearly; D – transparent to translucent; DE – 2.1-2.2; H – 2.5; CL – perfect; M – thin tabular crystals, scaly and pulverulent aggregates, earthy.
Origin and occurrence: Secondary, originated from the pyrite oxidation, together with other sulfates. Tabular crystals come from the Dexter No. 7 mine, Calf Mesa, Utah, USA and also from Železník , Slovakia. Fine crystals were found in many localities in Chile (Tierra Amarilla, Chuquicamata). It also occurs in Rammelsberg, Germany.

Copiapite, 96 mm, Copiapó, Chile

Serpierite, 80 mm, Příbram, Czech Republic

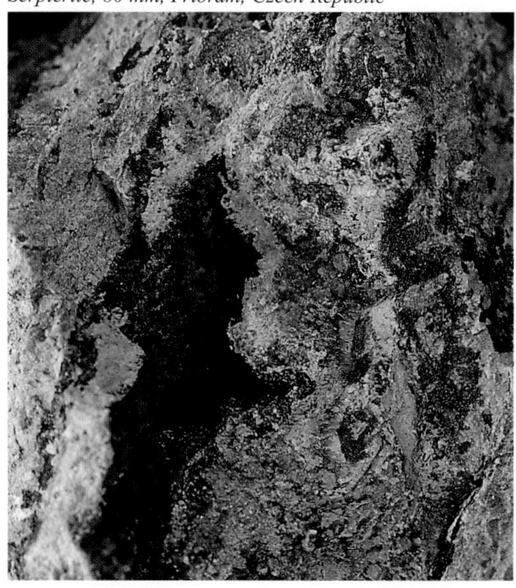

Devilline

$CaCu_4(SO_4)_2(OH)_6 \cdot 3\ H_2O$

MONOCLINIC ● ●

Properties: C – emerald-green; S – white to light green; L – vitreous to pearly; D – transparent to translucent; DE – 3.1; H – 2.5; CL – perfect; M – thin tabular pseudo-hexagonal crystals and coatings.
Origin and occurrence: Secondary, as a result of the oxidation of Cu sulfides, associated with other secondary Cu minerals. The world's best specimens come from Špania Dolina, Slovakia, where crystal rosettes up to 10 mm (3/8 in) in diameter were found in the past century; also known from Botallack, Cornwall, UK and Tsumeb, Namibia.

Serpierite

$Ca(Cu,Zn)_4(SO_4)_2(OH)_6 \cdot 3\ H_2O$

MONOCLINIC ● ●

Properties: C – blue; S – white; L – vitreous; D – transparent to translucent; DE – 3.1; H – 2.5; CL – perfect; M – tabular crystals, coatings.
Origin and occurrence: Secondary, originated in the oxidation zone of Cu-Zn deposits, together with smithsonite and other minerals. Its small crystals and aggregates occurred in Laurion, Greece. Recently confirmed at Příbram, Czech Republic.

Ettringite, 45 mm, Kuruman, South Africa

Sturmanite, 35 mm xx, Kuruman, South Africa

Ettringite

$Ca_6Al_2(SO_4)_3(OH)_{12} \cdot 26\ H_2O$

HEXAGONAL ● ●

Properties: C – colorless, yellow; S – white; L – vitreous to silky; D – transparent; DE – 1.8; H – 2.5; CL – perfect; M – prismatic crystals, fibrous aggregates.
Origin and occurrence: Metamorphic, associated with sturmanite. The world's best crystals reaching up to 100 mm (4 in) come from the N'Chwaning mine, Kuruman, South Africa. Crystals up to 4 mm (5/32 in) occurred in Franklin, New Jersey, USA. Also known from the contact metamorphic conditions in Crestmore, California, USA.

Sturmanite

$Ca_6Fe^{3+}{}_2(SO_4)_2[B(OH)_4], (OH)_{12} \cdot 25\ H_2O$

TRIGONAL ●

Properties: C – yellow, yellow-green; S – white; L – vitreous; D – transparent to translucent; DE – 1.9; H – 2.5; CL – perfect; M – flat dipyramidal crystals.
Origin and occurrence: : Probably metamorphic, associated with barite, hematite and ettringite. Crystals up to 140 mm (5½ in) long come from the N'Chwaning mine, Kuruman, South Africa.

Johannite

$Cu(UO_2)_2(SO_4)_2(OH)_2 \cdot 8\ H_2O$

TRICLINIC ● ●

Properties: C – green; S – light green; L – vitreous; D – transparent to translucent; DE – 3.3; H – 2-2.5; CL – good; M – prismatic to thick tabular crystals, scaly aggregates, coatings; R – radioactive.
Origin and occurrence: Secondary, as a result of the uraninite oxidation, together with other secondary U minerals. Crystals are known from Jáchymov, Czech Republic and Johanngeorgenstadt, Germany. It was common with zippeite in Central City, Colorado, USA and also reported from Mounana, Gabon.

Zippeite

$K_4(UO_2)_6(SO_4)_3(OH)_{10} \cdot 4\ H_2O$

ORTHORHOMBIC ● ● ●

Properties: C – orange-yellow; S – yellow; L – dull to earthy; D – opaque; DE – 3.7; H – not determined; CL – perfect; M – acicular and tabular crystals, pulverulent and acicular aggregates, coatings; LU – green; R – radioactive.
Origin and occurrence: Secondary, forming during the uraninite oxidation, associated with other secondary U minerals. Small tabular crystals come from Drmoul, Czech Republic. Coatings and

Johannite, 70 mm, Jáchymov, Czech Republic

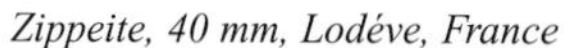

Zippeite, 40 mm, Lodéve, France

acicular aggregates are known from Jáchymov, Czech Republic and Central City, Colorado, USA; it also occurs in Shinkolobwe, Zaire.

Scheelite

$CaWO_4$

TETRAGONAL ● ● ● ●

Properties: C – colorless, gray-white, yellow-brown, orange, red, greenish; S – white; L – greasy to adamantine; D – translucent; DE – 6.1; H – 4.5-5; CL – good; F – conchoidal to uneven; M – pseudo-octahedral crystals, granular, massive; LU – blue-white.
Origin and occurrence: Magmatic in pegmatites, hydrothermal in greisens and metamorphic; parageneses vary significantly according to the origin. Beautiful brownish crystals up to 100 mm (4 in) in size come from Taewha and Tongwha in Korea. Similar crystals were recently found in China. Orange crystals up to 40 mm ($1^{9}/_{16}$ in) in size are associated with cassiterite crystals on quartz crystals from Iultin and Tenkergin, Russia. Clear crystals, weighing up to 50 kg (110 lb), occurred in pegmatites near Natas, Namibia. Crystals up to 70 mm (2¾ in) reported from several mines near Traversella, Italy. Beautiful red crystals up to 20 mm ($^{25}/_{32}$ in) were very rare in Príbram, Czech Republic. Yellow crystals up to 40 mm ($1^{9}/_{16}$ in) are known from quartz veins with pumpellyite in Obrí Dul, Czech Republic.
Application: W ore.

Scheelite, 77 mm, Sichuan, China

Stolzite, 20 mm, Arizona, U.S.A.

Crocoite, 44 mm, Dundas, Australia

Stolzite
$PbWO_4$

TETRAGONAL ● ●

Properties: C – gray-brown, orange-yellow, red, green; S – colorless; L – adamantine to resinous; D – transparent to translucent; DE – 7.9-8.3; H – 2.5-3; CL – imperfect; F – conchoidal to uneven; M – dipyramidal and thick tabular, striated crystals.
Origin and occurrence: Secondary, as a product of the oxidation of primary W minerals. Crystals up to 60 mm (2³/₈ in) from St Leger-de-Peyre, France. Prisms and needles up to 25 mm (1 in) long found in Broken Hill, New South Wales, Australia. Crystals up to 25 mm (1 in) also known from Tsumeb, Namibia. Crystals up to 20 mm (²⁵/₃₂ in) reported from Cínovec, Czech Republic and from the Black Pine mine, Montana, USA.

Crocoite
$PbCrO_4$

MONOCLINIC ● ●

Properties: C – orange, red; S – orange-yellow; L – adamantine to greasy; D – translucent; DE – 6; H – 3; CL – good; F – conchoidal to uneven; M – long prismatic to acicular crystals, crusts..
Origin and occurrence: Secondary, as a result of the galena oxidation in basic rocks. The world' best specimens come from the Dundas district, Tasmania, Australia where crystals up to 100 mm (4 in) long were found in several mines. Fine crystals up to 40 mm (1⁹/₁₆ in) long are known from Berezovsk, Ural mountains, Russia. Crystals up to 20 mm (²⁵/₃₂ in) were recently found in Callenberg, Germany.

Ferrimolybdite, 60 mm, Hůrky, Czech Republic

Betpakdalite, 70 mm, Bayan Tsogdo, Mongolia

Ferrimolybdite

$Fe_2(MoO_4)_3 \cdot 8 H_2O$

ORTHORHOMBIC ● ●

Properties: C – yellow, whitish; S – light yellow; L – adamantine, silky, earthy; D – opaque; DE – 4.4; H – 1-2; M – small acicular crystals, fibrous and radial aggregates, earthy.
Origin and occurrence: Secondary, as a product of the molybdenite oxidation. Microscopic crystals were found in Glen Innes, New South Wales, Australia. It occurred as yellow coatings in Climax and Telluride, Colorado, in the Getchell mine, Nevada, USA; also in Hurky near Cistá, Czech Republic.

Betpakdalite

$CaFe^{3+}H_8(MoO_4)_5(AsO_4)_2 \cdot 8 H_2O$

MONOCLINIC ●

Properties: C – lemon-yellow; S – yellow; L – vitreous, waxy, dull; D – opaque; DE – 3; H – 3; CL – good; M – short prismatic microscopic crystals, pulverulent.
Origin and occurrence: Secondary mineral. Originally described from Kara-Oba, Kazakhstan. It also occurred in Tsumeb, Namibia and in Krupka, Czech Republic.

Wulfenite

$PbMoO_4$

TETRAGONAL ● ● ●

Properties: C – yellow, orange, brownish, red, greenish; S – white; L – greasy to adamantine; D – transparent to translucent; DE – 6.3-7; H – 3; CL – good; F – uneven to conchoidal; M – thin tabular and dipyramidal crystals, granular, massive.
Origin and occurrence: Secondary, as a result of the galena oxidation, together with cerussite, vanadinite and other minerals. The best specimens are known from the Red Cloud mine near Yuma, Arizona, where red tabular crystals found up to 50 mm (2 in) diameter. Yellow-brown crystals up to 100 mm (4 in) come from the Glove mine, Arizona, USA. Thick tabular orange crystals up to 20 mm (25/32 in) in diameter found in the Erupcion mine, Villa Ahumada, Los Lamentos, Chihuahua. Beautiful yellow plates up to 60 mm (2 3/8 in) with orange mimetite spheres occurred in the San Francisco mine, Magdalena, Sonora, Mexico. Rare tabular crystals up to 70 mm (2¾ in) across come from Tsumeb, Namibia. Fine orange-yellow tabular crystals up to 20 mm (25/32 in) across, and pyramidal crystals are known from Bleiberg, Austria and Mežica , Slovenia. Crystals up to 100 mm (4 in) were found recently in Touissit, Morocco.

Wulfenite, 38 mm, Los Lamentos, Mexico

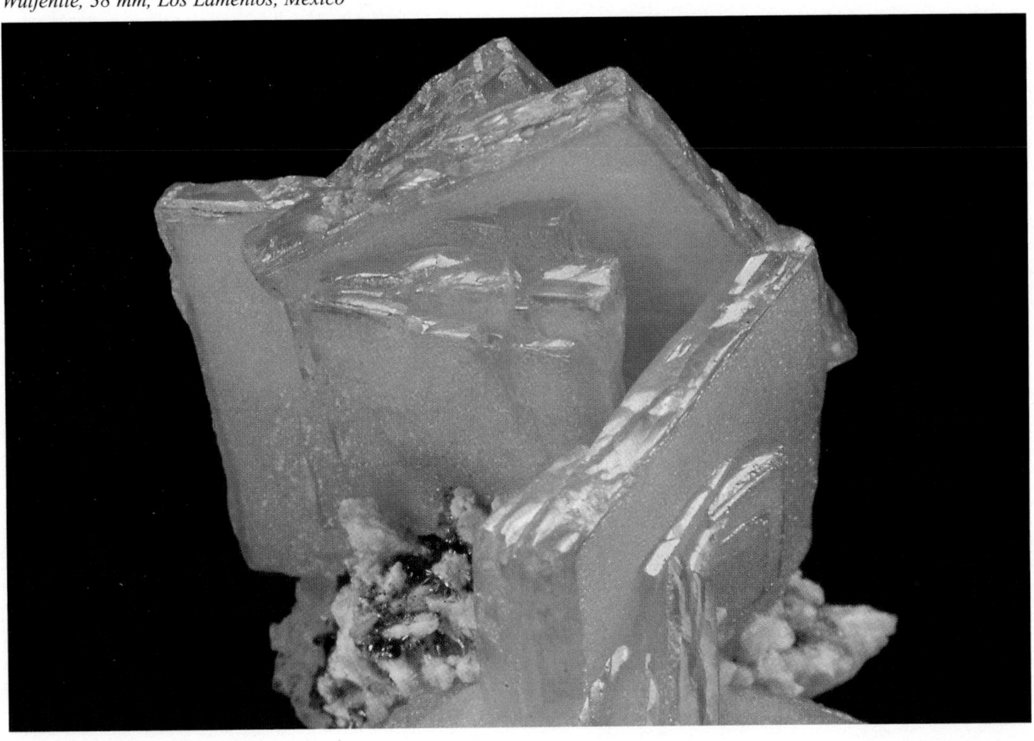

8. Phosphates

Lithiophosphate

Li_3PO_4

ORTHORHOMBIC ●

Properties: C – colorless, white, pinkish; S – white; L – vitreous; D – transparent to translucent; DE – 2.5; H – 4; CL – perfect; F – uneven; M – prismatic crystals, granular.
Origin and occurrence: Hydrothermal in granitic pegmatites, where it forms by replacement of spodumene and montebrasite. Cleavable aggregates up to 100 mm (4in) across occur in the Tanco Mine, Bernic Lake, Manitoba, Canada; crystals up to 25 mm (1 in) across in the Foote mine, Kings Mountain, North Carolina, USA.

Beryllonit

$NaBePO_4$

MONOCLINIC ● ●

Properties: C – colorless, white, yellowish; S – white; L – vitreous; D – transparent to translucent; DE – 2.8; H – 5.5-6; CL – perfect; F – uneven; M – short prismatic to tabular crystals, granular.
Origin and occurrence: Hydrothermal in cavities in granitic pegmatites where it is associated with herderite, albite and tourmaline. Crystals and their twins, up to 150 mm (6 in) across come from Stoneham and Newry, Maine, USA; also from Viitaniemi, Finland and Paprok, Afghanistan.

Adamite, 52 mm, Mapimi, Mexico
Lithiophosphate, 30 mm, Tanco, Canada

Beryllonite, 31 mm, Paprok, Afghanistan

Triphyllite

$LiFePO_4$

ORTHORHOMBIC ● ● ●

Properties: C – gray-green, gray-blue, gray, brown; S – gray-white; L – greasy to vitreous; D – transparent to translucent; DE – 3.4; H – 4-5; CL – good; F – uneven; M B short prismatic crystals, granular, massive.
Origin and occurrence: Magmatic in granitic pegmatites, associated with graftonite, sarcopside and many secondary phosphates. Large triphyllite crystals up to 1.5 m (5 ft) across are known from Hagendorf, Germany; from the Tip Top mine, Custer, South Dakota and Palermo No. 1 mine, North Groton, New Hampshire, USA; also from Hühnerkobel, Germany.

Triphyllite, 70 mm, Hagendorf, Germany

Lithiophilite, 120 mm, Viitaniemi, Finnland

Lithiophilite

$LiMnPO_4$

ORTHORHOMBIC ● ●

Properties: C – pink, red-brown, brown; S – white; L – vitreous; D – transparent to translucent; DE – 3.3; H – 4-5; CL – good; F – uneven; M – short prismatic crystals, granular, massive.
Origin and occurrence: Magmatic in granitic pegmatites. It is sometimes associated with triplite and triphyllite, typically replaced by many secondary phosphates. Large masses of lithiophillite, reaching up to 1 m (39³/₈ in), occur in Karibib, Namibia; Kitumbe, Rwanda. Other localities are Mangualde, Portugal; Tanco mine, Bernic Lake, Manitoba, Canada; Stewart Lithia Mine, Pala, California, USA.

Purpurite, 40 mm, Usakos, Namibia

Berzeliite, 120 mm, Langban, Sweden

Purpurite

$Mn^{3+}PO_4$

ORTHORHOMBIC ● ●

Properties: C – pink, purple, dark brown; S B red-purple; L – dull to velvet; D – translucent to opaque; DE – 3.7; H – 4-4.5; CL – good; F – uneven; M B granular, massive.
Origin and occurrence: Hydrothermal, as a product of lithiophillite replacement in granitic pegmatites. It is usually associated with many secondary phosphates. It is known from Kitumbe, Rwanda; Usakos and Sandamab, Namibia; the Tip Top and Bull Moose mines, Custer, South Dakota and Branchville, Connecticut, USA.

Berzeliite

$(Ca,Na)_3 (Mg,Mn)_2 (AsO_4)_3$

CUBIC ● ●

Properties: C – yellow, orange; S – red-purple; L – resinous; D – transparent to translucent; DE – 4.1; H – 4-4.5; CL – none; F – conchoidal to uneven; M – isometric crystals, granular, massive.
Origin and occurrence: Metamorphic, together with haussmanite, rhodonite and tephroite. It occurs as massive in Langban and Nordmark, Sweden.

Whitlockite

$Ca_9(Mg,Fe)H(PO_4)_7$

TRIGONAL ● ●

Properties: C – colorless, white, yellowish, pinkish; S – white; L – vitreous to dull; D – transparent to translucent; DE – 3.1; H – 5; CL – none; F – conchoidal to uneven; M – rhombohedral crystals, granular, massive.

Origin and occurrence: Hydrothermal, as a product of replacement of primary phosphates in granitic pegmatites, rare in sedimentary rocks – phosphorites; very rare magmatic in meteorites. Mainly associated with apatite and carbonates. Imperfect crystals around 10 mm (3/8 in) across, occur in the Palermo No. 1 mine, North Groton, New Hampshire; also in the Tip Top mine, Custer, South Dakota, USA.

Xenotime-(Y)

YPO_4

TETRAGONAL ● ● ●

Properties: C – brown, yellow, gray, greenish; S – white; L – vitreous to resinous; D – transparent, translucent to opaque; DE – 4.5; H – 4.5; CL – good; F – conchoidal to uneven; M – long prismatic to tabular crystals, granular; R – sometimes weakly radioactive and metamict.
Origin and occurrence: Magmatic in granitic and alkaline pegmatites, granites and syenites; hydrothermal in the Alpine-type veins; metamorphic in gneisses; common in placers. It is associated with monazite-(Ce) and zircon. Perfect prismatic crystals up to 100 mm (4in) across occur mainly in pegmatites in Kragerö and Hitterö, Norway; Ytterby, Sweden; in several places in Madagascar; in Ichikawa, Japan. Crystals about 20 mm (25/32 in) are known from the Alpine-type veins in Binntal, Switzerland.
Application: ore of rare earth elements.

Whitlockite, 10 mm xx, New Hampshire, U.S.A.

Xenotime-(Y), 14 mm x, Washington Co., U.S.A.

Monazite-(Ce)
$CePO_4$

MONOCLINIC ● ● ●

Properties: C – yellow, brown, red-brown, orange, gray-green; S – white; L – vitreous to resinous; D – transparent, translucent to opaque; DE – 4.6; H – 5-5.5; CL – good; F – conchoidal to uneven; M – long prismatic to tabular crystals, granular; R – sometimes weakly radioactive and metamict.
Origin and occurrence: Magmatic in granitic and alkaline pegmatites, granites, syenites, carbonatites; hydrothermal in the Alpine-type veins and greisens; metamorphic in gneisses; common in placers. It is associated with apatite, xenotime-(Y) and zircon. Perfect prismatic crystals up to 200 mm (7 7/8 in) come from Mars Hill, North Carolina, Trout Creek Pass, Colorado, USA; also from Arendal, Norway; Ambatofotsikely and Ampangabé, Madagascar, where masses weighing several kg are common; crystals, up to 200 mm (7 7/8 in) across found in Minas Gerais, Jaguaraçu, Brazil.
Application: ore of rare earth elements.

Monazite-(Ce), 31 mm, Felicio, Brazil

Hydroxylherderite
$CaBe(PO_4)(OH,F)$

MONOCLINIC ● ●

Properties: C – colorless, white, yellowish, greenish; S – white; L – vitreous to dull; D – transparent to translucent; DE – 3.0; H – 5-5.5; CL – good; F – conchoidal to uneven; M – prismatic to tabular crystals, radial aggregates, granular.
Origin and occurrence: Hydrothermal in cavities in granitic pegmatites and in greisens. Perfect crystals, up to 120 mm (4 11/16 in) long come from Marilaca and together with colored tourmalines from Virgem da Lapa, Minas Gerais, Brazil; also known from Topsham and Stoneham, Maine, USA.

Amblygonite
$LiAl(PO_4)(F,OH)$

TRICLINIC ● ● ●

Properties: C – colorless, white, yellowish, bluish, gray; S – white; L – vitreous to dull; D – transparent to translucent; DE – 3.1; H – 5.5-6; CL – good; F – conchoidal to uneven; M – short prismatic crystals, granular, massive.

Hydroxylherderite, 22 mm, Anza, California, U.S.A.

Montebrasite, 80 mm, White Picacho, Arizona, U.S.A.

Wagnerite, 18 mm x, Werfen, Austria

Origin and occurrence: Magmatic in granitic pegmatites and some granites; rare hydrothermal in greisens and in ore veins. Large amblygonite masses, several meters across, come from pegmatites, like the Beecher, Custer and Hugo mines, Keystone, South Dakota, USA, where its blocks weighed up to 200 tons; also known from Viitaniemi, Finland; Utö, Sweden. Typical representative of the quartz-amblygonite veins is Vernérov, Czech Republic.
Application: Li ore and raw material for ceramics.

Origin and occurrence: Magmatic in granitic pegmatites; metamorphic in gneisses and eclogites; hydrothermal in quartz veins and in salt deposits. The most famous finds come from the quartz veins in Höllgraben and Radelgraben, the Alps, Austria, where wagnerite occur together with lazulite and forms crystals up to 30 mm ($1^{3}/_{16}$ in) across; also known from Mangualde, Portugal and Bodenmais, Germany.

Montebrasite

$LiAl(PO_4)(OH,F)$

TRICLINIC ● ● ●

Properties: C – colorless, white, yellowish, yellow, bluish, gray; S – white; L – vitreous to dull; D – transparent to translucent; DE – 3.0; H – 5.5-6; CL – good; F – conchoidal to uneven; M – short prismatic to tabular crystals, granular, massive.
Origin and occurrence: Magmatic in granitic pegmatites and granites; hydrothermal in granitic pegmatites. It occurs very frequently with amblygonite, which it replaces. Large masses are known from Montebras, France; also from the Tin Mountain mine, Custer, South Dakota, USA. Perfect tabular crystals up to 150 mm (6 in) across are found in Taquaral, Minas Gerais, Brazil.

Wagnerite

$Mg_2(PO_4)(F,OH)$

MONOCLINIC ● ● ●

Properties: C – light yellow, yellow-green, yellow-brown, green; S – white; L – vitreous to greasy; D – transparent, translucent to opaque; DE – 3.2; H – 5-5.5; CL – imperfect; F – conchoidal to uneven; M – short prismatic crystals, granular, massive.

Amblygonite, 55 mm, Minas Gerais, Brazil

Zwieselite, 30 mm grain, Dolní Bory, Czech Republic

Zwieselite

$Fe_2(PO_4)(F,OH)$

MONOCLINIC ● ● ●

Properties: C – dark brown to black-brown; S – light brown; L – vitreous to greasy; D – translucent to opaque; DE – 4.0; H – 5-5.5; CL – imperfect; F – conchoidal to uneven; M – short prismatic crystals, granular, massive.
Origin and occurrence: Magmatic in granitic pegmatites; rare hydrothermal in greisens. It is associated with apatite, triplite and secondary phosphates. Imperfect crystals and granular aggregates are known from pegmatites near Zwiesel, Germany and Dolní Bory, Czech Republic.

Libethenite, 80 mm, L'ubietová, Slovakia

Triplite, 70 mm, Viitaniemi, Finnland

Triplite

$Mn_2(PO_4)(F,OH)$

MONOCLINIC ● ● ●

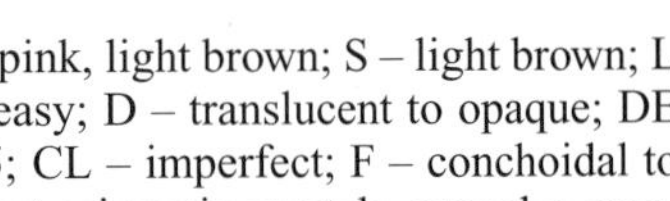

Properties: C – pink, light brown; S – light brown; L – vitreous to greasy; D – translucent to opaque; DE – 3.8; H – 5-5.5; CL – imperfect; F – conchoidal to uneven; M – short prismatic crystals, granular, massive.
Origin and occurrence: Magmatic in granitic pegmatites and granites; hydrothermal in greisens and in quartz veins. It is usually hydrothermally altered and replaced by secondary phosphates. Huge masses several meters in size come from pegmatites in Karibib, Namibia; it also occurs in Mangualde, Portugal and in Sukula, Finland.

Libethenit

$Cu_2(PO_4)(OH)$

ORTHORHOMBIC ● ● ●

Properties: C – black-green to light green; S – olive-green; L – greasy; D – translucent; DE – 3.9; H – 4; CL – imperfect; F – conchoidal to uneven; M – short prismatic and dipyramidal crystals, botryoidal aggregates, granular, massive.
Origin and occurrence: Secondary in Cu deposits, where it occurs together with malachite, pseudomalachite and brochantite. Perfect crystals, up to 30 mm ($1^{3}/_{16}$ in) across, come from the Rokana mine, Zambia; crystals, up to 10 mm ($^{3}/_{8}$ in), are known from Kambove, Zaire; L'ubietová, Slovakia; and Nizhniy Tagil, Ural Mts., Russia.

Olivenite, 80 mm, Tsumeb, Namibia

Olivenite
$Cu_2(AsO_4)(OH)$

ORTHORHOMBIC ● ● ●

Properties: C – olive-green, green-brown, gray-green to gray; S – light green; L – greasy; D – translucent to opaque; DE – 4.4; H – 3; CL – imperfect; F – conchoidal to uneven; M – long to short prismatic and dipyramidal crystals, acicular and radial aggregates, massive.
Origin and occurrence: Secondary in Cu deposits. It is associated with other secondary Cu minerals, malachite, azurite and scorodite. Perfect crystals up to 10 mm (3/8 in) across occurred in Wheal Gorland, Cornwall, and Tavistock, Devon UK; also in Tsumeb, Namibia and Ashburton Downs, Western Australia.

Adamine
$Zn_2(AsO_4)(OH)$

ORTHORHOMBIC ● ● ●

Properties: C – yellow-green, yellow, green, colorless, purple; S – white; L – vitreous; D – transparent to translucent; DE – 4.4; H – 3.5; CL – imperfect; F – conchoidal to uneven; M – long to short prismatic and dipyramidal crystals, acicular and radial aggregates, massive; LU – yellow-green.

Origin and occurrence: Secondary in Zn deposits, associated with hemimorphite, goethitee and smithsonite. Rich druses of green and rare purple crystals up to 70 mm (2¾ in) across come from Mina Ojuela, Mapimi, Durango, Mexico; also known from Tsumeb, Namibia; Laurion, Greece; and Cap Garonne, France.

Adamite, 34 mm, Mapimi, Mexico

Lazulite, 11 mm x, Lincoln Co., U.S.A.

Lazulite

$MgAl_2(PO_4)_2(OH)_2$

MONOCLINIC ● ● ●

Properties: C – dark to light blue, blue-green; S – white; L – vitreous; D – transparent to translucent; DE – 3.1; H – 5.5-6; CL – imperfect; F – conchoidal to uneven; M – short prismatic and dipyramidal crystals, granular, massive.
Origin and occurrence: Hydrothermal in quartz veins and granitic pegmatites, where it is formed by decomposition of primary phosphates; metamorphic in quartzites. Imperfect crystals up to 100 mm (4 in) across come from Horrsjöberg, Sweden; perfect crystals about 50 mm found in Ashudi, Pakistan; also known from Big Fish River, Yukon, Canada and near Werfen, Austria.

Scorzalite

$FeAl_2(PO_4)_2(OH)_2$

MONOCLINIC ● ● ●

Properties: C – dark blue, blue-green; S – white; L – vitreous; D – transparent to translucent; DE – 3.3; H – 6; CL – imperfect; F – conchoidal to uneven; M – prismatic crystals, granular, massive.
Origin and occurrence: Hydrothermal in granitic pegmatites, as a replacement product of primary phosphates, rare in quartz veins. Dark blue, granular aggregates up to 100 mm (4in) across occur in the Palermo No. 1 and No. 2 mines, North Groton, New Hampshire and the Victory mine, Custer, South Dakota, USA.

Rockbridgeite

$Fe^{2+}Fe^{3+}_4(PO_4)_3(OH)_5$

ORTHORHOMBIC 

Properties: C – dark and light green, black-green; S – green; L – dull; D – translucent to opaque; DE – 3.4; H – 4.5; CL – good; F – uneven; M – acicular crystals, radial aggregates and crusts, granular, massive.
Origin and occurrence: Secondary in granitic pegmatites and in Fe deposits. It mostly originates from the hydrothermal alteration of primary phosphates, mainly triphyllite, and it is associated with other secondary phosphates. Rich radial aggregates up to 50 mm (2 in) across come from Hagendorf, Germany; the Tip Top mine, Custer, South Dakota and the Fletcher mine, Groton, New Hampshire, USA.

Scorzalite, 80 mm, Palermo No.2 Mine, U.S.A.

Rockbridgeite, 70 mm, Hagendorf, Germany

Frondelite, 15 mm xx, Custer, U.S.A.

Frondelite

$MnFe^{3+}{}_4(PO_4)_3(OH)_5$

ORTHORHOMBIC ● ● ●

Properties: C – light olive-green, brown, black-green; S – green; L – dull; D – translucent to opaque; DE – 3.5; H – 4.5; CL – good; F – uneven; M – acicular crystals, radial aggregates and crusts, granular, massive.

Origin and occurrence: Secondary in granitic pegmatites, where it forms as a result of the hydrothermal alteration of primary phosphates, mainly lithiophillite. Radial aggregates occur in the Fletcher mine, Groton, New Hampshire, USA; also in Sapucaia, Minas Gerais, Brazil.

Dufrenite

$Fe^{2+}Fe^{3+}{}_4(PO_4)_3(OH)_5 \cdot 2\,H_2O$

MONOCLINIC ● ● ●

Properties: C – dark green, black-green, black; S – green; L – vitreous to dull; D – translucent to opaque; DE – 3.4; H – 3.5-4.5; CL – good; F – uneven; M – radial aggregates and crusts, granular, massive.

Origin and occurrence: Secondary in granitic pegmatites and in the oxidation zone of Fe deposits; it forms in pegmatites as a result of the hydrothermal alteration of primary phosphates. Radial aggregates occur in pegmatites in Hagendorf and Hühnerkobel, Germany; also in gossan in the Wheal Phoenix mine, Cornwall, UK.

Dufrenite, 9 mm, Lancaster, U.S.A.

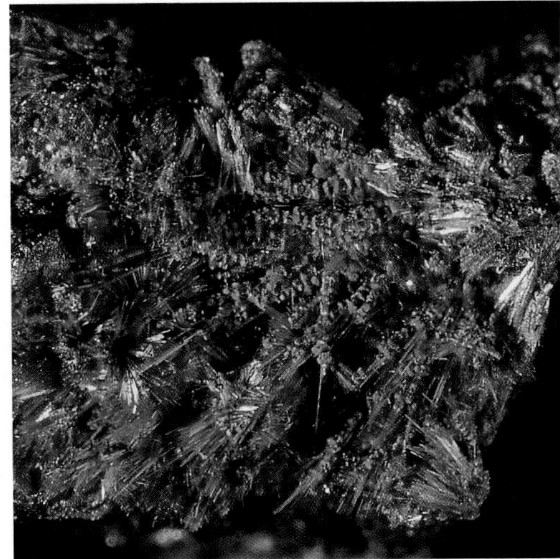

Pseudomalachite

$Cu_5(PO_4)_2(OH)_4$

MONOCLINIC 

Properties: C – green, black-green; S – green; L – vitreous to dull; D – translucent to opaque; DE – 4.3; H – 4.5-5; CL – good; F – uneven; M – short prismatic crystals, botryoidal aggregates with radial structure, massive.

Origin and occurrence: Secondary in Cu deposits, associated with malachite, chrysocolla, libethenite and goethite. Beautiful botryoidal aggregates and large masses come from Nizhniy Tagil, Ural mountains, Russia; also from Ehl and Virneberg, Germany; from L'ubietová, Slovakia; and also from many localities in Shaba province, Zaire.

Pseudomalachite, 80 mm, L'ubietová, Slovakia

Cornubite, 30 mm, Farbište, Slovakia

Cornubite

$Cu_5(AsO_4)_2(OH)_4$

TRICLINIC ●

Properties: C – light to dark green; S – light green; L – vitreous to dull; D – translucent; DE – 4.8; H – not determined; CL – not determined; F – uneven; M – botryoidal aggregates, massive.

Origin and occurrence: Secondary in Cu deposits. It was found together with olivenite and clinoclase in the Bedford United quarry, Cornwall, UK; also in Ashburton Downs, Western Australia and Farbište, Slovakia.

Cornetite, 10 mm aggregate, Lumumbashi, Zair

Augelite, 34 mm, Yukon Territory, Canada

Augelite

$Al_2(PO_4)(OH)_3$

MONOCLINIC ● ● ●

Properties: C – colorless, white, yellowish, pinkish; S- white; L – vitreous to dull; D – transparent to translucent; DE – 2.7; H- 4.5-5; CL – good; F – uneven; M – thick tabular to prismatic crystals, granular, massive.

Origin and occurrence: Hydrothermal in granitic pegmatites as a product of primary phosphates replacement; rare metamorphic in quartzites. Massive aggregates several decimeters across occur in pegmatites in Burango, Rwanda; also in the Hugo mine, Custer, South Dakota and Mount White, California, USA. Crystals up to 20 mm ($^{25}/_{32}$ in) across come from Rapid Creek, Yukon, Canada and from the Champion mine, California; rare small crystals are also found in the Palermo No. 1 mine, North Groton, New Hampshire, USA.

Cornetite

$Cu_3(PO_4)(OH)_3$

ORTHORHOMBIC ●

Properties: C – dark blue to blue-green; S – light blue; L – vitreous; D – transparent to translucent; DE – 4.1; H – 4.5; CL – none; F – conchoidal to uneven; M – short prismatic crystals, coatings.

Origin and occurrence: Secondary in Cu deposits. It is rare in the Etoile mine near Lumumbashi and in Kalagi, Shaba province, Zaire; also occurs in Yerington, Nevada and in Saginaw Hill, Arizona, USA.

Clinoclase, 40 mm, Cornwall, UK

Conichalcite, 70 mm, Tintic, U.S.A.

Clinoclase

$Cu_3(AsO_4)(OH)_3$

MONOCLINIC ● ●

Properties: C – dark green-blue to black-green; S – blue-green; L – vitreous; D – transparent to translucent; DE – 4.4; H – 2.5-3; CL – perfect; F – uneven; M – prismatic and tabular crystals, botryoidal aggregates.
Origin and occurrence: Secondary in Cu deposits, often associated with malachite, azurite and other secondary Cu minerals. Spherical aggregates up to 10 mm ($^3/_8$ in) known from the Majuba Hill mine, Nevada, USA; occurs rarely near Tavistock, Devon, UK; also in Novoveská Huta, Slovakia.

Conichalcite

$CuCa(AsO_4)(OH)$

ORTHORHOMBIC ● ● ●

Properties: C – yellow-green to emerald-green; S – light green; L – vitreous to greasy; D – transparent to translucent; DE – 4.3; H – 4.5; CL – none; F – uneven; M – short prismatic crystals, botryoidal aggregates with radial structure, massive.
Origin and occurrence: Secondary in Cu deposits. It occurs as rich botryoidal aggregates in Otavi, Namibia; Tintic, Utah, in the Higgins mine, Bisbee, Arizona and Yerington, Nevada, USA.

Duftite

$PbCu(AsO_4)(OH)$

ORTHORHOMBIC ● ●

Properties: C – yellow-green, olive-green to gray-green; S – light green; L – vitreous to greasy; D – translucent; DE – 6.5; H – 3; CL – not determined; F – uneven; M – small crystals, botryoidal aggregates and coatings.
Origin and occurrence: Secondary in base metals deposits. It is associated with malachite and azurite in Tsumeb, Namibia; in Mina Ojuela, Mapimi, Durango, Mexico; and in Moldava, Czech Republic.

Duftite, 90 mm, Tsumeb, Namibia

Descloizite, 38 mm, Grootfontein, Namibia

Descloizite

$PbZn(VO_4)(OH)$

ORTHORHOMBIC ● ● ●

Properties: C – red-orange, red-brown to brown-black, gray-green; S – light yellow-brown; L – greasy; D – transparent, translucent to opaque; DE – 6.1; H – 3-3.5; CL – none; F – conchoidal to uneven; M – crystals of different habits, mostly pyramidal or prismatic, botryoidal and skeletal aggregates, massive.
Origin and occurrence: Secondary in base metals deposits. It is mainly associated with pyromorphite, mimetite, vanadinite and other secondary Pb minerals. Occurs as crystals up to 30 mm ($1^{3}/_{16}$ in) long in Tsumeb and Berg Aukas, Namibia; also Broken Hill, Zambia; and the Mammoth mine, Tiger, Arizona, USA.

Arsendescloizite

$PbZn(AsO_4)(OH)$

ORTHORHOMBIC ●

Properties: C – light yellow; S – white; L – adamantine to greasy; D – transparent to translucent; DE – 6.1; H – 4; CL – none; F – conchoidal to uneven; M – tabular crystals, rosette-like aggregates.

Arsendescloizite, 95 mm, Mapimi, Mexico

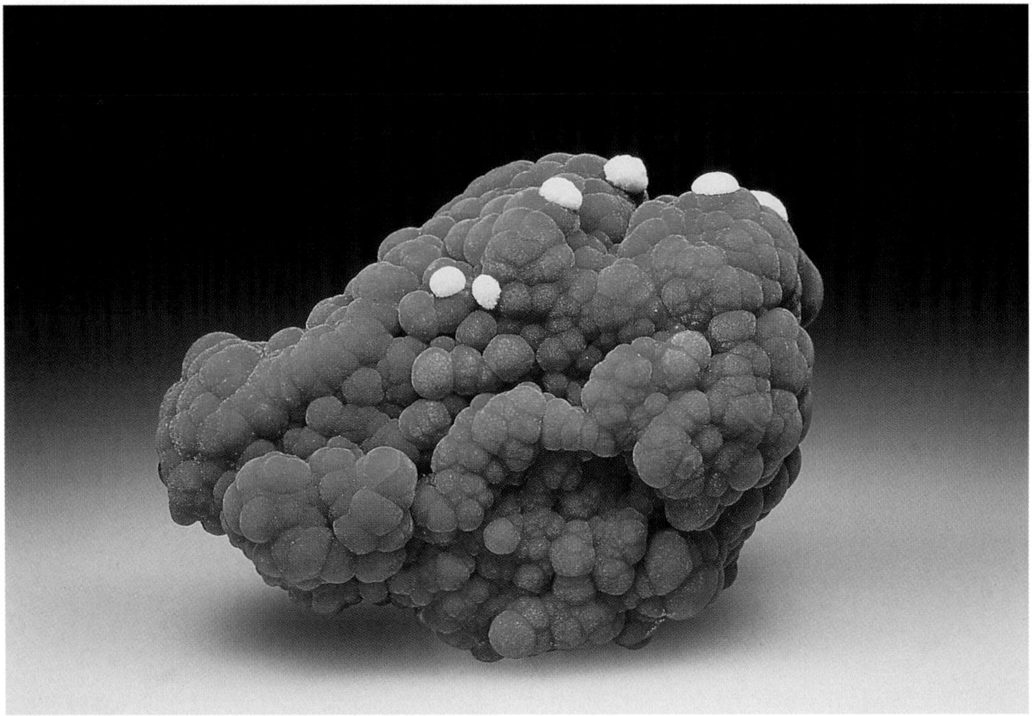

Origin and occurrence: Secondary in base metals deposits, associated with mimetite and goethite. Found rarely as crystals of 1 mm ($^1/_{32}$ in) in size in Tsumeb, Namibia; also Mina Ojuela, Mapimi, Durango, Mexico.

Mottramite
$PbCu(AsO_4)(OH)$

ORTHORHOMBIC ● ●

Properties: C – grass-green to black-green; S – light green; L – vitreous to dull; D – transparent to opaque; DE – 5.9; H – 3-3.5; CL – none; F – conchoidal to uneven; M – crystals of different habits, botryoidal and dendritic aggregates, crusts and coatings, massive.
Origin and occurrence: Secondary in base metals deposits, associated with mimetite, descloizite and vanadinite. It occurs in Mottram, Cheshire, UK; in Tsumeb, Namibia; and Mammoth mine, Tiger, Arizona, USA.

Brazilianite
$NaAl_3(PO_4)_2(OH)_4$

MONOCLINIC ● ● ●

Properties: C – colorless, white, yellowish, yellow-green; S – white; L – vitreous; D – transparent to translucent; DE – 3.0; H – 5.5; CL – good; F – uneven; M – short prismatic to isometric crystals, radial aggregates, granular.
Origin and occurrence: Hydrothermal in cavities in granitic pegmatites, where it is associated with fluorapatite, albite and tourmaline. It occasionally originates as a product of amblygonite replacement. Perfect yellow-green crystals up to 150 mm (6 in) across found in cavities in pegmatites in Conselheira Pena and Córrego Frio, Linopolis, Minas Gerais and from Pietras Lavradas, Paraíba, Brazil.

Mottramite, 60 mm, Tiger, Arizona, U.S.A.

Cafarsite, 9 mm x, Binntal, Switzerland

Cafarsite
$Ca_8(Ti,Fe^{2+},Fe^{3+},Mn)(As^{3+}O_3)_{12} \cdot 4\ H_2O$

CUBIC ●

Properties: C – dark brown; S – yellow-brown; L – submetallic; D – translucent; DE – 3.9; H – 5.5-6; CL – none; F – conchoidal; M – isometric crystals.
Origin and occurrence: Hydrothermal along cracks in Alpine-type veins. Cubic crystals up to 30 mm ($1^3/_{16}$ in) across come from Binntal, Switzerland and Pizzo Cervandone, Italy.

Brazilianite, 47 mm, Conselheira Pena, Brazil

Carminite

$PbFe^{3+}_2(AsO_4)_2(OH)_2$

ORTHORHOMBIC ● ●

Properties: C – crimson-red, red-brown; S – red-yellow; L – adamantine to pearly; D – translucent; DE – 5.5; H – 3.5; CL – good; F – conchoidal to uneven; M – prismatic crystals, acicular, radial, felt-like to porous aggregates, coatings and crusts.
Origin and occurrence: Secondary in base metals deposits, associated with mimetite and scorodite. Crystals up to 10 mm (3/8 in) across occur in Tsumeb, Namibia; in Mina Ojuela, Mapimi, Durango, Mexico; in Calstock, Cornwall, UK.

Carminite, 2 mm xx, Nadap, Hungary

Bayldonite, 34 mm, Tsumeb, Namibia

Bayldonite

$PbCu_3(AsO_4)_2(OH)_2 \cdot H_2O$

MONOCLINIC ● ●

Properties: C – yellow-green, olive-green; S – light green; L – resinous; D – translucent; DE – 5.5; H – 4.5; CL – not determined; F – uneven; M – earthy aggregates, coatings, massive.
Origin and occurrence: Secondary in the oxidation zone in hydrothermal Cu deposits and in greisens. It occurs as common yellow-green coatings and crystals up to 10 mm (3/8 in) across, associated with large azurite and mimetite crystals in Tsumeb, Namibia; also as coatings in St. Day, Cornwall, UK; and in Moldava, Czech Republic.

Vésignéite

$BaCu_3(VO_4)_2(OH)_2$

MONOCLINIC ● ●

Properties: C – yellow-green, olive-green; S – light yellow-green; L – vitreous to dull; D – translucent;

Vésignéite, 80 mm, Vrančice, Czech Republic

DE – 4.1; H – 3-4; CL – good; F – uneven; M – earthy and pulverulent aggregates and coatings, massive.

Origin and occurrence: Secondary in base metals deposits and in sediments, containing Cu sulfides. It occurs as yellow-green coatings in sediments in Horní Kalná, Czech Republic; also in the Mashamba West mine, Zaire; and together with barite and psilomelane in Friedrichsroda, Germany.

Arsentsumebite

$Pb_2Cu(AsO_4)(SO_4)(OH)$

MONOCLINIC ●

Properties: C – bluish-green to light green; S – light green; L – dull; D – translucent; DE – 6.5; H – 3; CL – good; F – uneven; M – earthy aggregates, coatings, massive.

Origin and occurrence: Secondary in the oxidation zone, where it forms as a result of mimetite replacement. Found rarely in Moldava, Czech Republic; and in Tsumeb, Namibia, where it forms pseudo-morphs after azurite crystals.

Arsentsumebite, 40 mm, Moldava, Czech Republic

Fluorapatite, 50 mm, Dusso, Pakistan

Fluorapatite
$Ca_5(PO_4)_3F$

HEXAGONAL ● ● ● ● ●

Properties: C – colorless, white, yellowish, pinkish, blue, purple, green, brown with various hues; S – white; L – vitreous to dull; D – transparent to translucent, sometimes opaque; DE – 3.2; H – 5; CL – imperfect; F – conchoidal to uneven; M – long prismatic to tabular crystals, botryoidal, earthy and fibrous aggregates, massive; LU – yellow.
Origin and occurrence: Magmatic in granites, syenites, diorites, gabbros and various types of pegmatites, also in volcanic rocks; hydrothermal in quartz veins, ore veins, greisens and Alpine-type veins; metamorphic in different types of gneisses, migmatites, mica schists, skarns and amphibolites; in different types of sedimentary rocks. Perfect, short prismatic, purple transparent crystals up to 40 x 40 mm ($1^{9}/_{16}$ x $1^{9}/_{16}$ in) are renowned from pegmatite in the Pulsifer quarry, Mount Apatite, Auburn, Maine, USA; pink crystals from Dusso, Pakistan. Perfect crystals about 100 mm (4in) across also from Alpine-type veins, e.g. in Fiesch, Switzerland; also known from the quartz veins in greisens, associated with wolframite, in Panasqueira, Portugal; in Horní Slavkov, Czech Republic; Ehrenfriedersdorf, Germany. Also found in skarns at Cerro de Mercado, Durango, Mexico. Large deposits of mas-

Chlorapatite, 20 mm x, Bob's Lake, Ontario, Canada

sive apatite located in Kola Peninsula, Russia; crystals weighing up to 300 kg (660 lb) from the vicinity of Clear Lake, Ontario, Canada.
Application: main source of P, chemical industry, fertilizer.

Chlorapatite
$Ca_5(PO_4)_3Cl$

HEXAGONAL ● ● ●

Properties: C – white, various hues of yellow; S – white; L – vitreous to dull; D – transparent to translucent; DE – 3.2; H – 5; CL – imperfect; F – conchoidal to uneven; M – long prismatic to tabular crystals, granular.
Origin and occurrence: Magmatic in nepheline syenites and their pegmatites, some gabbros and volcanic rocks, also in meteorites; metamorphic in skarns. It is usually associated with scapolite, amphibole, titanite and magnetite. Perfect prismatic crystals up to 35 cm ($13^{12}/_{16}$ in) long, come from pegmatites in Bamle, Norway; also from Kurokura, Japan.

Hydroxylapatite, 2 mm xx, Tornaszentandrás, Hungary

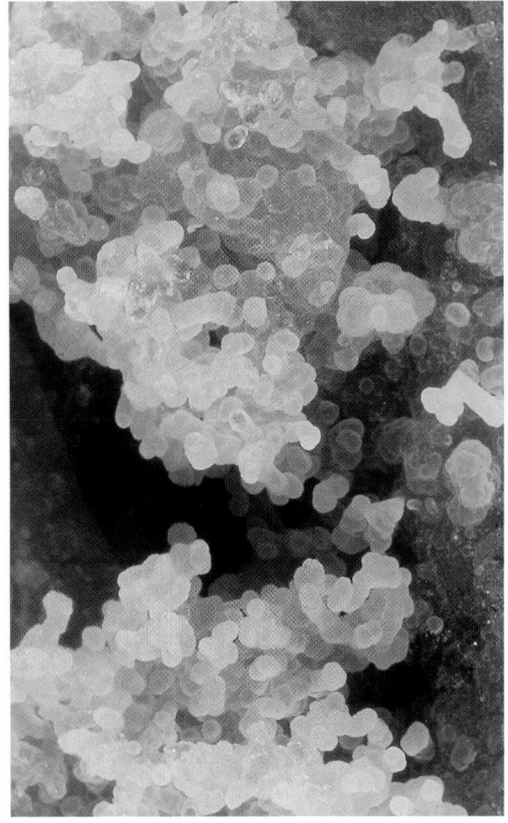

Hydroxylapatite
$Ca_5(PO_4)_3(OH)$

HEXAGONAL ● ● ●

Properties: C – white, yellow, various hues of gray; S – white; L – vitreous to dull; D – transparent to translucent; DE – 3.2; H – 5; CL – imperfect; F – conchoidal to uneven; M – short prismatic to tabular crystals, acicular aggregates, granular.
Origin and occurrence: Metamorphic in talc schists and serpentinites; hydrothermal in granitic pegmatites; sedimentary in organic remnants. Crystals up to 30 mm ($1^{3}/_{16}$ in) across known from Snarum, Norway; Hospental, Switzerland; and Eagle, Colorado, USA.

Carbonate-fluorapatite
$Ca_5(PO_4,CO_3)_3F$

HEXAGONAL ● ● ●

Properties: C – white, gray; S – white; L – vitreous to dull; D – transparent to translucent; DE – 3.2; H – 5; CL – imperfect; F – conchoidal to uneven; M B spherical and botryoidal aggregates, massive.
Origin and occurrence: Hydrothermal in ore veins and along the cracks in volcanic rocks. Rich botryoidal aggregates are known together with hyalite opal from Valec, Czech Republic and the Wheal Franco mine, Tavistock, Devon, UK.

Carbonate-fluorapatite, 80 mm, Staffel, Germany

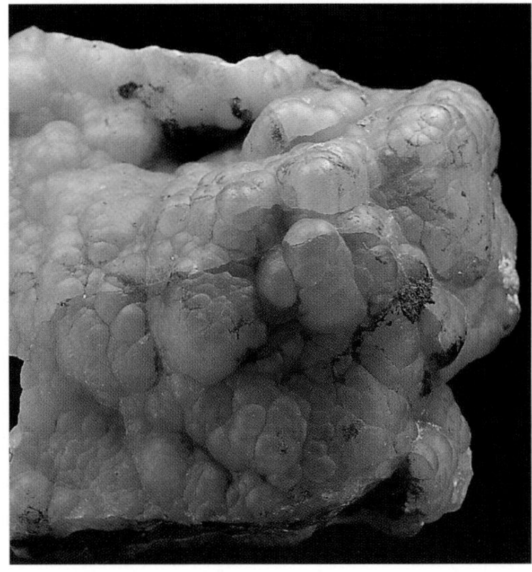

Pyromorphite, 41 mm, Les Farges, France

Pyromorphite

$Pb_5(PO_4)_3Cl$

HEXAGONAL ● ● ● ●

Properties: C – green, brown, yellow, orange, white, gray; S -white; L – adamantine to greasy; D – transparent to translucent; DE – 7.1; H – 3.5-4; CL – imperfect; F – conchoidal to uneven; M – long prismatic to tabular and pyramidal crystals, botryoidal aggregates with radial structure, acicular and earthy aggregates, massive.

Origin and occurrence: Secondary in the oxidation zone of Pb deposit, associated with cerussite, goethitee and other secondary minerals. Perfect crystals up to 40 mm ($1^{9}/_{16}$ in) long come from many localities, e.g. green and brown ones from Bad Ems and Zschopau, Germany; yellow-brown ones from Les Farges, France; green, brown and yellow ones from Mina Ojuela, Mapimi, Durango, Mexico. Green, orange and brown crystals, up to 60 mm ($2^{3}/_{8}$ in) long, are known from the Bunker Hill mine, Idaho, USA.

Pyromorphite, 62 mm, Les Farges, France

Mimetite, 14 mm x, Johanngeorgenstadt, Germany

Mimetite

$Pb_5(AsO_4)_3Cl$

HEXAGONAL ● ● ●

Varieties: campylite

Properties: C – yellow, orange, brown, yellow-brown, white, gray; S – white; L – adamantine to greasy; D – transparent to translucent; DE – 7.3; H – 3.5-4; CL – imperfect; F – conchoidal to uneven; M – long to short prismatic and pyramidal crystals, botryoidal aggregates with radial structure, acicular and earthy aggregates, granular.
Origin and occurrence: Secondary in Pb deposits, associated with pyromorphite and goethite. Perfect crystals up to 20 mm ($^{25}/_{32}$ in) long come from Johanngeorgenstadt, Germany; campylite occurs in Drygill, UK; crystals up to 50 mm (2 in) across found in Tsumeb, Namibia; Santa Eulalia and San Pedro, Chihuahua, Mexico. Beautiful yellow crystals up to 30 mm ($1^{3}/_{16}$ in) found recently in Hat Yai province, Thailand.

Vanadinite

$Pb_5(VO_4)_3Cl$

HEXAGONAL ● ● ●

Properties: C – yellow, orange, red, brown, yellow-brown; S – white; L – adamantine to greasy; D – transparent to translucent; DE – 6.9; H – 2.5-3; CL – none; F – conchoidal to uneven; M – long to short prismatic and pyramidal crystals, botryoidal aggregates with radial structure, acicular and earthy aggregates, granular.
Origin and occurrence: Secondary in Pb deposits, associate with pyromorphite, wulfenite and goethite. Perfect crystals up to 130 mm ($5^{2}/_{16}$ in) long come from Djebel Mahseur and Mibladen, Morocco. Prismatic crystals were also found in Tsumeb, Namibia; in the Old Yuma, Red Cloud, Apache and Mammoth mines, Arizona, USA.

Vanadinite, 64 mm, Mibladen, Morocco

Atelestite, 70 mm, Schneeberg, Germany

Atelestite

$Bi_2(AsO_4)O(OH)$

MONOCLINIC ●

Properties: C – yellow, yellow-green; S – white; L – adamantine to greasy; D – transparent to translucent; DE – 7.0; H – 4.5-5; CL – imperfect; F – conchoidal to uneven; M – small tabular crystals, spherical aggregates.
Origin and occurrence: Secondary in the oxidation zone of Bi deposits. It was found in Schneeberg, Germany.

Huréaulite

$Mn_5(PO_4)_2[PO_3(OH)]_2 . 4 H_2O$

MONOCLINIC ● ● ●

Properties: C – orange-red, pink, purplish, white, gray; S – white; L – vitreous; D – transparent to translucent; DE – 3.2; H – 3.5; CL – good; F – uneven; M – prismatic to tabular crystals, coatings, granular, massive.

Huréaulite, 53 mm, Shingus, Pakistan

Variscite, 30 mm, Fairfield, U.S.A.

Origin and occurrence: Secondary in granitic pegmatites, as a result of alteration of the primary phosphates, mostly lithiophillite, triphyllite and graftonite. Purplish crystals up to 50 mm (2 in) across come from Shingus, Pakistan. Similar crystals found in Hagendorf, Germany. Also known from Mangualde, Portugal; Sao Jose da Safira, Minas Gerais, Brazil; and from the Tip Top mine, Custer, South Dakota, USA.

Variscite

$Al(PO_4) . 2 H_2O$

ORTHORHOMBIC ● ● ●

Properties: C – colorless, greenish, blue-green; S – white; L – vitreous; D – transparent to translucent; DE – 2.6; H – 3.5-4.5; CL – good; F – uneven; M – isometric crystals, botryoidal aggregates, nodules, coatings, massive.
Origin and occurrence: : Hydrothermal in cracks in sedimentary rocks, rich in Al and P, also in phosphates deposits. It is associated with apatite, wavellite and other phosphates. Renowned greenish nodules up to 30 cm (12 in) in diameter come from Clay Canyon, Fairfield, Utah, USA; also known from Ronneburg, Germany and Jivina near Beroun, Czech Republic.

Strengite

$Fe^{3+}(PO_4) . 2 H_2O$

ORTHORHOMBIC ● ● ●

Properties: C – colorless, pinkish, red-purple; S – white; L – vitreous; D – transparent to translucent; DE – 2.8; H – 3.5-4.5; CL – good; F – conchoidal; M – isometric, tabular to short prismatic crystals, botryoidal aggregates with radial structure, coatings.
Origin and occurrence: Secondary in granitic

Strengite, 34 mm, Svappavaara, Sweden

Scorodite, 10 mm x, Zacatecas, Mexico

pegmatites where it forms as a result of hydrothermal replacement of primary phosphates; in the oxidation zone of Fe deposit together with goethite. Purple crystals up to 5 mm ($^{3}/_{16}$ in) across, come from the Bull Moose mine, Custer, South Dakota, USA; also known from Pleystein, Germany and from Teškov, Czech Republic.

Scorodite

$Fe^{3+}(AsO_4) \cdot 2\ H_2O$

ORTHORHOMBIC ● ● ●

Properties: B – light green, gray-green, olive-green, colorless, blue, yellow-brown; S – light green; L – vitreous to resinous; D – transparent to translucent; DE – 3.3; H – 3.5-4; CL – imperfect; F – conchoidal; M – dipyramidal to short prismatic crystals, botryoidal and earthy aggregates, coatings, granular, massive.

Origin and occurrence: Secondary in the oxidation zone of ore deposits, associated with arsenopyrite, löllingite and other arsenides; in granitic pegmatites; hydrothermal in the hot springs. Perfect crystals up to 50 mm (2 in) across come from Tsumeb, Namibia; the Kiura mine, Oita, Japan; Mina Ojuela, Mapimi, Durango, Mexico; green crusts, several cm thick, were found in Djebel Debar, Algeria.

Phosphophyllite, 18 mm x, Potosí, Bolivia

Ludlamite, 36 mm, Morococala, Bolivia

Phosphophyllite

$Zn_2Fe(PO_4)_2 \cdot 4\ H_2O$

MONOCLINIC ● ●

Properties: C – colorless, blue-green, blue; S – white; L – vitreous; D – transparent to translucent; DE – 3.1; H – 3-3.5; CL – good; F – uneven; M – long to short prismatic and thick tabular crystals, granular.
Origin and occurrence: Secondary in the oxidation zone of ore deposits; in granitic pegmatites, where it replaces primary phosphates. Perfect crystals, up to 140 mm (5½ in) across come from sulfide cavities in the Unificada mine, Potosí, Bolivia. Also occurs at Hagendorf, Germany.

Ludlamite

$Fe_3(PO_4)_2 \cdot 4\ H_2O$

MONOCLINIC ● ●

Properties: C – light green, green; S – white; L – vitreous; D – transparent to translucent; DE – 3.2; H – 3.5; CL – perfect; F – uneven; M – thin to thick tabular crystals, granular, massive.
Origin and occurrence: Secondary in the oxidation zone of ore deposits; in granitic pegmatites, where it replaces primary phosphates. Perfect crystals up to 90 mm (3⁹⁄₁₆ in) across known from the San Antonio mine, Santa Eulalia, Chihuahua, Mexico; also from Morococala, Bolivia and the Blackbird district, Idaho, USA.

Anapaite

$Ca_2Fe(PO_4)_2 \cdot 4\ H_2O$

TRICLINIC ● ●

Properties: C – light to dark green, colorless; S – white; L – vitreous; D – transparent to translucent; DE – 2.8; H – 3.5; CL – perfect; F – uneven; M – thin to thick tabular crystals, rosette-like aggregates, granular.
Origin and occurrence: Secondary in the oxidation zone of Fe deposits and in sedimentary rocks, rich in P. Tabular crystals several mm across and spherical aggregates, up to 30 mm (1³⁄₁₆ in) across, come from the cracks within oolitic ores near Anapa and Kerch, Crimea, Ukraine; also known from Bellaver de Cerdena, Spain.

Vivianite

$Fe_3(PO_4)_2 \cdot 8\ H_2O$

MONOCLINIC ● ● ● ●

Properties: C – colorless when fresh, quickly oxidizes to blue, green, purple, black-blue; S – white to bluish; L – vitreous; D – transparent, translucent to opaque; DE – 2.7; H – 1.5-2; CL – perfect; F – uneven; M – long prismatic to acicular crystals, fibrous, earthy to pulverulent aggregates, coatings, granular, massive.
Origin and occurrence: Secondary in the oxidation zone of Fe deposits; in granitic pegmatites, where it forms by the replacement of primary phosphates and in sedimentary rocks in proximity of organic

Anapaite, 18 mm aggregates, Anapa, Ukraine

material; hydrothermal in the ore deposits. Crystals up to 1.5 m (5 ft) long found in clay sediments in Anloua, Cameroon; crystals up to 200 mm (7⁷/₈ in) across known from Morococala, Bolivia. Smaller crystals come from Trepca, Serbia; Leadville, Colorado, Bingham Canyon, Utah, USA; also from Kerch and Anapa, Crimea, Ukraine.

Erythrine

$Co_3(AsO_4)_2 . 8\ H_2O$

MONOCLINIC ● ● ●

Properties: C – dark purple, pink, colorless; S – light pink to white; L – vitreous; D – transparent to translucent; DE – 3.2; H – 1.5-2.5; CL – perfect; F – uneven; M – long prismatic, acicular to tabular crystals, earthy aggregates, coatings, granular, massive.

Origin and occurrence: Secondary in the oxidation zone of Co, Ni and U deposits. Tabular crystals up to 60 mm (2³/₈ in) long come from Bou Azzer, Morocco. Other important localities are Schneeberg, Germany; Talmessi, Iran; Cobalt, Ontario, Canada; and Mount Cobalt, Queensland, Australia.

Erythrine, 29 mm, Mexico

Vivianite, 48 mm, Morococala, Bolivia

Annabergite, 55 mm, Laurion, Greece

Annabergite

$Ni_3(AsO_4)_2 . 8 H_2O$

MONOCLINIC ● ● ●

Varieties: cabrerite (Mg contents)

Properties: C – light to dark green, white; S – white; L – vitreous; D – transparent to translucent; DE – 3.2; H – 1.5-2.5; CL – perfect; F – uneven; M – long prismatic to acicular and tabular crystals, acicular and earthy aggregates, coatings, granular, massive.
Origin and occurrence: Secondary in the oxidation zone of Ni deposits, associated with erythrite. Crystals up to 5 mm (3/16 in) across occur in Gukurören, Turkey; Sierra Cabrera, Spain; and Laurion, Greece. Nodules up to 20 mm (25/32 in) come from the Snowbird mine, Montana, USA.

Symplesite

$Fe_3(AsO_4)_2 . 8 H_2O$

TRICLINIC ● ●

Properties: C – bluish, dark blue, light green, black-green; S – white; L – vitreous; D – transparent to translucent; DE – 3.0 H – 2.5; CL – perfect; F – uneven; M – acicular to tabular crystals, spherical aggregates with radial structure, earthy aggregates and coatings, granular, massive.
Origin and occurrence: Secondary in the oxidation zone of ore deposits and in granitic pegmatites with

Symplesite, 12 mm xx, Baia Sprie, Romania

Picropharmacolite, 128 mm, France

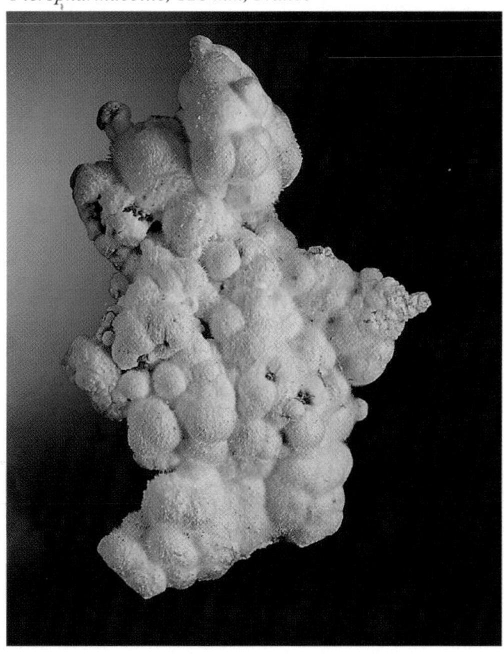

As minerals, mainly arsenopyrite and löllingite. It occurs in Schneeberg, Germany; Baia Sprie, Romania and Trebsko, Czech Republic.

Picropharmacolite

$Ca_4MgH_2(AsO_4)_4 . 12 H_2O$

TRICLINIC ● ●

Properties: C – colorless, white; S – white; L – vitreous; D – transparent to translucent; DE – 2.6; H – 2.5; CL – perfect; F – uneven; M – acicular crystals, spherical aggregates with radial structure, coatings.
Origin and occurrence: Secondary in the oxidation zone of ore deposits with As minerals, mostly arsenopyrite and löllingite. It is also known from Ste-Marie-aux-Mines and Salsigne, France; Freiberg, Germany; and Jáchymov, Czech Republic.

Brushite

$CaH(PO_4) . 2 H_2O$

MONOCLINIC ● ●

Properties: C – colorless, white; S – white; L – vitreous, pearly on cleavage planes; D – transparent to translucent; DE – 2.3; H – 2.5; CL – perfect; F – uneven; M – prismatic, acicular to tabular crystals, earthy aggregates and coatings, massive.
Origin and occurrence: Secondary on bat and bird excrements and bones, it impregnates bones, along the cracks of phosphorites. Tabular crystals up to 20 mm ($^{25}/_{32}$ in) across come from Quercy, France also occurs near Oran, Algeria and Pig Hole, Virginia, USA.

Brushite, 4 mm xx, Domica, Slovakia

Legrandite, 25 mm xx, Mapimi, Mexico

Legrandite

$Zn_2(AsO_4)(OH) . H_2O$

MONOCLINIC ● ●

Properties: C – colorless, yellow, purple; S – white; L – vitreous; D – transparent to translucent; DE – 4.0 H – 4.5; CL – imperfect; F – uneven; M – long prismatic crystals and their inter-growths, radial aggregates.
Origin and occurrence: Secondary in the oxidation zone of Zn deposits and in pegmatites. Prismatic crystals up to 250 mm ($9^{13}/_{16}$ in) long found in Mina Ojuela, Mapimi, Durango, Mexico. It is also known from Galiléia, Minas Gerais, Brazil and Tsumeb, Namibia.

Euchroite, 10 mm xx, L'ubietová, Slovakia

Euchroite

$Cu_2(AsO_4)(OH) . 3 H_2O$

ORTHORHOMBIC ● ●

Properties: C – emerald-green; S – white; L – vitreous; D – transparent to translucent; DE – 3.5; H – 3.5-4; CL – imperfect; F – conchoidal to uneven; M – short prismatic to thick tabular crystals, massive.

Origin and occurrence: Secondary in the oxidation zone of Cu deposits, associated with olivenite and malachite. Thick tabular crystals up to 30 mm (1³/₁₆ in) come from L'ubietová, Slovakia. It also occurs in Zapacica, Bulgaria and Chessy, France.

Vauxite, 20 mm xx, Llallagua, Bolivia

Strunzite, 10 mm xx, Hagendorf, Germany

Vauxite

$FeAl_2(PO_4)_2(OH)_2 . 6 H_2O$

TRICLINIC ● ●

Properties: C – light to dark blue; S – white; L – vitreous; D – transparent to translucent; DE – 2.4; H – 3.5; CL – none; F – uneven; M – tabular crystals, radial aggregates, massive.

Origin and occurrence: Secondary in the oxidation zone of Sn deposits, associated with wavellite. Crystals occur in the Siglo XX Mine, Llallagua, Bolivia.

Strunzite

$MnFe^{3+}{}_2(PO_4)_2(OH)_2 . 6 H_2O$

MONOCLINIC ● ● ●

Properties: C – light yellow, yellow-brown; S – yellowish; L – vitreous; D – transparent to translucent; DE – 2.5; H – not determined; CL – not determined; F – uneven; M – acicular crystals, acicular and fibrous aggregates.

Origin and occurrence: Secondary in granitic pegmatites, as a result of weathering of primary phosphates, mostly triphyllite; rarely hydrothermal in the cracks of Fe-rich sedimentary rocks. Acicular crystals up to 20 mm (²⁵/₃₂ in) across known from Hagendorf, Germany; the Palermo No. 1 and No. 2 mines, North Groton and the Fletcher mine, Groton, New Hampshire, USA.

Cacoxenite, 1 mm xx, Hellertown, Pennsylvania, U.S.A.

Cacoxenite

$(Fe^{3+},Al)_{25}(PO_4)_{17}O_6(OH)_{12} . 75\ H_2O$

HEXAGONAL ● ● ●

Properties: C – light yellow, yellow-brown, orange; S – yellow; L – silky; D – transparent to translucent; DE – 2.3; H – 3-4; CL – not determined; F – uneven; M – acicular and fibrous aggregates, often with radial structure, botryoidal crusts and coatings.
Origin and occurrence: Hydrothermal on the cracks of sedimentary Fe ores, associated with wavellite; rare as secondary in granitic pegmatites, as a product of weathering of primary phosphates. Crystals up to 10 mm (3/8 in) across come from the Horcajo mine, Ciudad Real, Spain; golden-yellow acicular aggregates occur in Hrbek near Svatá Dobrotivá and Trenice, Czech Republic and in Amberg, Germany.

Beraunite

$Fe^{2+}Fe^{3+}{}_5(PO_4)_4(OH)_5 . 4\ H_2O$

MONOCLINIC ● ●

Properties: C – red-brown, red, gray-green; S – yellow to green-brown; L – vitreous to dull; D – translucent; DE – 3.0; H – 3.5-4; CL – good; F – uneven; M – acicular aggregates, often with radial structure, botryoidal crusts and coatings.
Origin and occurrence: Hydrothermal on the cracks of sedimentary Fe ores, typically together with wavellite; secondary in granitic pegmatites, as a product of weathering of primary phosphates. Acicular aggregates up to 10 mm (3/8 in) across occur in Mount Indian, Alabama, USA; also known from Hrbek near Svatá Dobrotivá, Czech Republic; and Amberg, Germany.

Beraunite, 3 mm aggregates, Nekészény, Hungary

Diadochite

$Fe^{3+}{}_2(PO_4)(SO_4)(OH) . 5\ H_2O$

TRICLINIC ● ● ●

Properties: C – yellow-brown, brown, red-brown, yellow-green, gray-green; S – yellow to light brown; L – dull, waxy; D – translucent to opaque; DE – 2.0-2.4; H – 3; CL – not determined; F – uneven, conchoidal, earthy; M – nodules, coatings and crusts, massive.
Origin and occurrence: Secondary in the oxidation zone of Fe deposits. Diadochite caves in abandoned mines are known from Saalfeld, Germany; nodules are found in New Idria, California and Eureka, Nevada, USA.

Diadochite, 40 mm, Récsk, Hungary

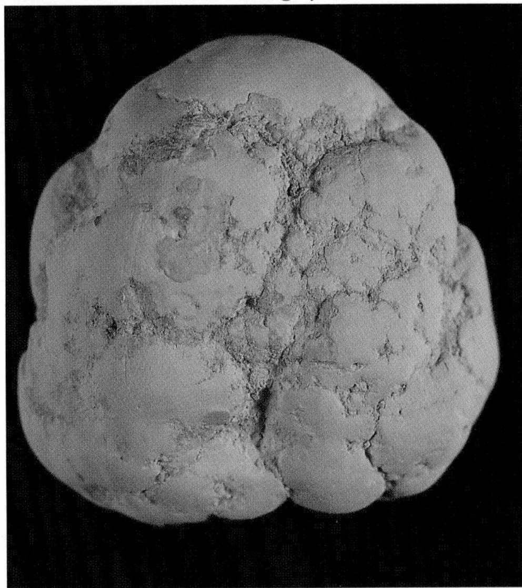

Wavellite, 40 mm, Třenice, Czech Republic

Wavellite

$Al_3(PO_4)_2(OH,F)_3 \cdot 5\ H_2O$

ORTHORHOMBIC ● ● ●

Properties: C – colorless, white, greenish, light blue-green, green, yellowish; S – white; L – vitreous to pearly; D – transparent to translucent; DE – 2.4; H – 3.5-4; CL – perfect; F – uneven; M – isometric crystals, hemispherical aggregates with radial structure, botryoidal aggregates, nodules, coatings, massive.
Origin and occurrence: Hydrothermal in the cracks of Al and P-rich sediments, also in phosphate deposits, ore veins and pegmatites. Beautiful hemispherical aggregates up to 40 mm (1 9/16 in) in diameter occur in Pencil, Garland and Magnet Cove, Arkansas, USA; also known from Trenice and Mílina, Czech Republic and Ronneburg, Germany.

Eosphorite, 4 mm xx, Lavra da Ilha, Brazil

Eosphorite

$MnAl(PO_4)(OH)_2 \cdot H_2O$

ORTHORHOMBIC ● ●

Properties: C – pinkish, colorless, white, brownish, red-brown; S – white; L – vitreous to pearly; D – transparent to translucent; DE – 3.1; H – 5; CL – imperfect; F – uneven to conchoidal; M – long to short prismatic crystals, radial aggregates, granular.
Origin and occurrence: Secondary in granitic pegmatites, as a product of hydrothermal replacement of primary phosphates. Crystals up to 100 mm (4in) long found in the Joao Modesto dos Santos mine, Minas Gerais, Brazil. It also occurs in Rapid Creek, Yukon, Canada.

Turquoise

$CuAl_6(PO_4)_4(OH)_8 \cdot 4\ H_2O$

TRICLINIC ● ● ●

Properties: C – blue, blue-green, green; S – white; L – waxy; D – transparent, translucent to opaque; DE – 2.9; H – 5-6; CL – good; F – conchoidal to uneven; M – short prismatic crystals, botryoidal aggregates, coatings, massive.
Origin and occurrence: Secondary in the surface parts of rocks with elevated contents of P and Cu, e.g. in the oxidation zone of some Cu deposits. Small crystals occurred near Lynch Station, Virginia, USA. Massive blue and blue-green concretions come from Mount Ali Mirsai near Maden, Iran. Other localities are Cortez, Nevada; Los Cerillos and Eureka, New Mexico; and Bisbee, Arizona, USA.
Application: popular gemstone.

Turquoise, 3 mm aggregates, Humboldt Co., U.S.A.

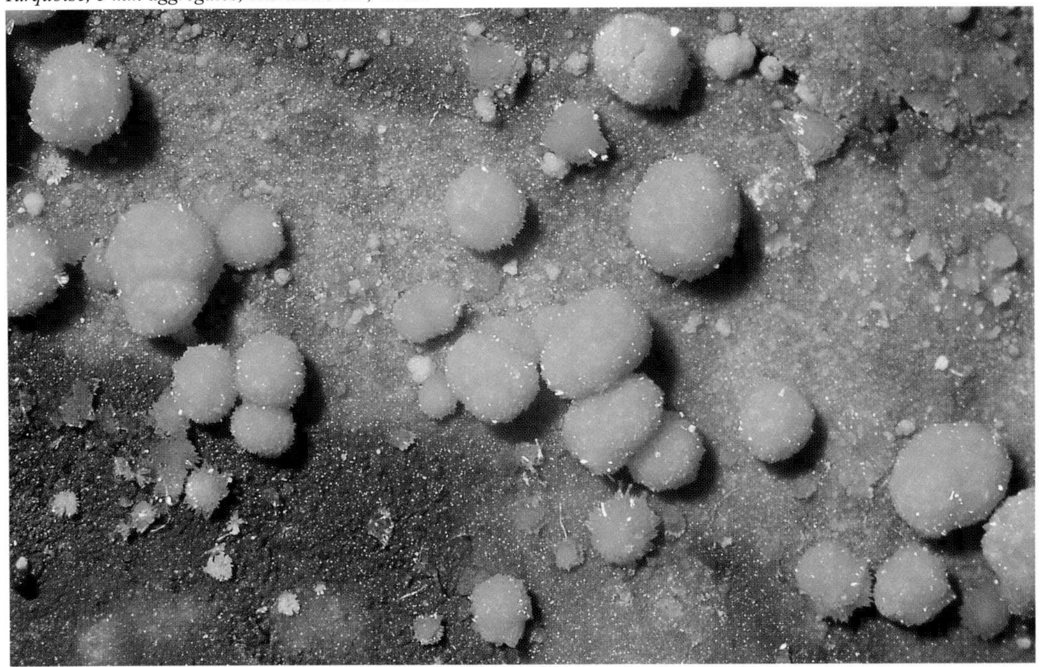

Chalcosiderite

$CuFe^{3+}{}_6(PO_4)_4(OH)_8 \cdot 4\ H_2O$

TRICLINIC ● ●

Properties: C – dark green; S – white; L – vitreous; D – transparent to translucent; DE – 3.3; H – 4.5; CL – good; F – conchoidal to uneven; M – short prismatic crystals, coatings.

Origin and occurrence: Secondary in the oxidation zone of some Cu deposits, together with goethite, dufrenite and pharmacosiderite. It occurs in Bisbee, Arizona, USA; in the Wheal Phoenix mine, Cornwall, UK; Schneckenstein, Germany; and Horní Slavkov, Czech Republic.

Turquoise, 50 mm, Kazakhstan

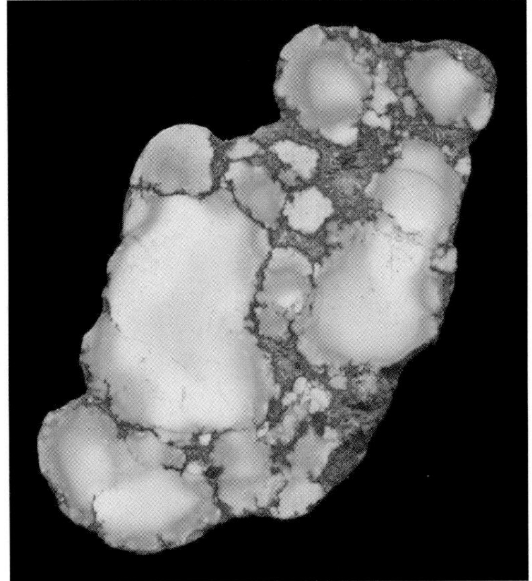

Chalcosiderite, 90 mm, Cornwall, UK

Chenevixite, 100 mm, Chuquicamata, Chile

Tyrolite, 90 mm, Novoveská Huta, Slovakia

Chenevixite

$Cu_2Fe^{3+}{}_2(AsO_4)_2(OH)_4 . H_2O$

MONOCLINIC ●

Properties: C – dark green, olive-green to yellow-green; S – yellow-green; L – greasy; D – translucent; DE – 3.9; H – 3.5-4.5; CL – not determined; F – conchoidal to uneven; M – earthy aggregates, coatings, massive.

Origin and occurrence: Secondary in the oxidation zone of Cu deposits, associated with malachite, tyrolite, azurite and other minerals. Massive aggregates occur in the Mammoth mine, Tintic, Utah, USA; also in Klein Spitzkopje, Namibia.

Tyrolite

$CaCu_5(AsO_4)_2(CO_3)(OH)_4 . 6\ H_2O$

ORTHORHOMBIC ● ● ●

Properties: C – apple-green, green-blue to blue; S – light green to blue-green; L – vitreous to pearly; D – transparent to translucent; DE – 3.3; H – 2; CL – perfect; F – uneven; M B scaly and fan-shaped aggregates, coatings and crusts.

Origin and occurrence: Secondary in the oxidation zone of Cu deposits, frequently associated with chalcophyllite. Rich aggregates occur in the Majuba Hill mine, Nevada, also in Tintic, Utah, USA. It is also known from Brixlegg, Austria; Saalfeld and Schneeberg, Germany; and Novoveská Huta, Slovakia.

Delvauxite

$CaFe^{3+}{}_4(PO_4,SO_4)(OH)_8 . 4\text{-}6\ H_2O$

AMORPHOUS ● ● ●

Properties: C – yellow-brown, brown, red-brown, black-brown; S – yellow; L – greasy, waxy; D – translucent to opaque; DE – 1.8-2.0; H – 2.5; CL – not determined; F – conchoidal, earthy; M – nodules, stalactites, coatings and crusts, massive.

Origin and occurrence: Secondary in oxidation zone of Fe deposits. Nodules of 50 cm (20 in) across in Czech Republic. Also known in Berneau and Richelle, Belgium; Zelezník, Slovakia; Kerch, Crimea, Ukraine.

Delvauxite, 80 mm, Nučice, Czech Republic

Bukovskyite, 40 mm, Kaňk, Czech Republic

Bukovskyite

$Fe^{3+}{}_2(AsO_4)(SO_4)(OH) . 7 H_2O$

TRICLINIC ●

Properties: C – yellow-green, gray-green; S – yellowish white; L – dull to earthy; D – translucent to opaque; DE – 2.3; H – not determined; CL – not determined; F – earthy; M B botryoidal aggregates and nodules.
Origin and occurrence: Secondary on the old mine dumps where it forms as a product of arsenopyrite weathering. Nodules up to 60 cm (24 in), occur on medieval dumps in Kank near Kutná Hora, Czech Republic.

Chalcophyllite, 36 mm, Chile

Veszelyite

$(Cu,Zn)_3(PO_4)(OH)_3 . 2 H_2O$

MONOCLINIC ● ●

Properties: C – green, blue-green, dark blue; S – green; L – vitreous; D – translucent; DE – 3.4; H – 3.5-4; CL – good; F – uneven; M – short prismatic to tabular crystals, granular.
Origin and occurrence: Secondary in the oxidation zone of Cu-Zn deposits. Crystals up to 50 mm (2 in) across found in the Black Pine mine, Philipsburg, Montana, USA; also from Moravita, Romania; Arakawa, Japan; and Wanlockhead, Scotland, UK.

Chalcophyllite

$Cu_{18}Al_2(AsO_4)_3(SO_4)_3(OH)_{27} . 36 H_2O$

TRIGONAL ● ● ●

Properties: C – emerald-green, blue-green; S – light green; L – vitreous, pearly; D – transparent to translucent; DE – 2.6; H – 2; CL – perfect; F – uneven; M – tabular crystals, scaly, fan-shaped aggregates, coatings, massive.
Origin and occurrence: Secondary in the oxidation zone of Cu deposits, usually associated with tyrolite. Rich aggregates occur in the Majuba Hill, Nevada and in the Tintic district, Utah, USA. Nice specimens also come from Novoveská Huta and Piesky, Slovakia; Nizhniy Tagil, Ural mountains, Russia; and Cap Garrone, France.

Veszelyite, 13 mm, Philipsburg, U.S.A.

Liroconite, 86 mm, Cornwall, UK

Liroconite

$Cu_2Al(AsO_4)(OH)_4 . 4 H_2O$

MONOCLINIC ● ●

Properties: C – blue, green; S – light blue; L – vitreous to resinous; D – transparent to translucent; DE – 3.0; H – 2-2.5; CL – imperfect; F – conchoidal to uneven; M – lenticular, dipyramidal crystals, massive.

Origin and occurrence: Secondary in the oxidation zone of Cu deposits, together with olivenite, malachite and azurite. Perfect crystals up to 30 mm (1 3/16 in) across come from Redruth and St. Day, Cornwall, UK. It is also known from Cerro Gordo, California, USA.

Evansite, 5 mm aggregates, Sirk – Železník, Slovakia

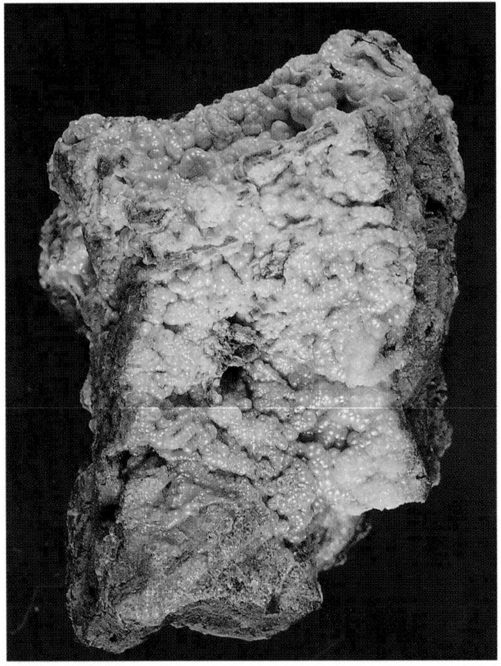

Evansite

$Al_3(PO_4)(OH)_6 . 6 H_2O$

AMORPHOUS 3

Properties: C – colorless, white, greenish, light blue-green, yellowish; S – white; L – vitreous, resinous, waxy; D – transparent to translucent; DE – 1.8-2.2; H – 3-4; CL – none; F – conchoidal; M – botryoidal, stalactitic and hemispherical aggregates and coatings.

Origin and occurrence: Secondary in the oxidation zone of Fe deposits, rich in P, associated with alophane and goethite. Rich stalactitic aggregates come from Zelezník, Slovakia and from Épernay, France.

Whiteite-(CaFeMg)

$CaFeMg_2Al_2(PO_4)_4(OH)_2 . 8 H_2O$

MONOCLINIC ● ●

Properties: C – brown; S – white; L – vitreous; D – transparent to translucent; DE – 2.6; H – 4; CL – good; F – uneven; M – short prismatic crystals.

Origin and occurrence: Hydrothermal in the cracks of P-rich sediments. Perfect crystals up to 20 mm (25/32 in) across come from Big Fish River, Yukon, Canada. Crystals up to 5 mm (3/16 in), are known from Lavra da Ilha de Taquaral, Minas Gerais, Brazil.

Whiteite-(CaFeMg), 65 mm, Yukon Territory, Canada

Jahnsite-(CaMnMg), 60 mm, Custer, U.S.A.

Wardite, 48 mm, Yukon Territory, Canada

Jahnsite-(CaMnMg)

$CaMnMg_2Fe^{3+}{}_2(PO_4)_4(OH)_4 \cdot 8\ H_2O$

MONOCLINIC ● ●

Properties: C – yellow, light to dark brown; S – light yellow; L – vitreous; D – transparent to translucent; DE – 2.6; H – not determined; CL – good; F – uneven; M – short to long prismatic crystals, granular.
Origin and occurrence: Secondary in granitic pegmatites, where it forms as a result of replacement of primary phosphates. Perfect crystals up to 10 mm (3/8 in) across occur in the Tip Top mine, Custer, South Dakota, USA. It is also known from Hagendorf, Germany.

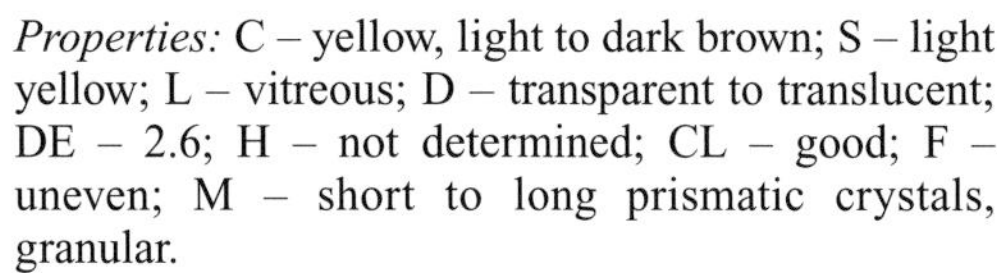

Wardite

$NaAl_3(PO_4)_2(OH)_4 \cdot 2\ H_2O$

TETRAGONAL ● ● ●

Properties: C – colorless, white, greenish, light blue-green; S -white; L – vitreous; D – transparent to translucent; DE – 2.8; H – 5; CL – perfect; F – uneven; M – dipyramidal crystals, radial and hemispherical aggregates, coatings, crusts, granular.
Origin and occurrence: Secondary in granitic pegmatites, where it forms as a product of primary phosphate replacement; hydrothermal in the cracks in P-rich sediments. Perfect crystals up to 30 mm (1 3/16 in) across come from Rapid Creek, Yukon, Canada. It is also known from Piedras Lavradas, Paraíba, Brazil.

Cyrilovite

$NaFe^{3+}{}_3(PO_4)_2(OH)_4 \cdot 2\ H_2O$

TETRAGONAL ● ●

Properties: C – yellow, orange, brown-yellow; S – yellow; L – vitreous; D – transparent to translucent; DE – 3.1; H – 4; CL – good; F – conchoidal; M – tabular and dipyramidal crystals, coatings and crusts.
Origin and occurrence: Secondary in granitic pegmatites, where it forms as a result of replacement of primary phosphates. Small crystals occur in Cyrilov, Czech Republic; Hagendorf, Germany; and Sapucaia, Minas Gerais, Brazil.

Cyrilovite, 60 mm, Iron Monarch, Australia

Pharmacosiderite, 35 mm, Cornwall, UK

Arseniosiderite, 3 mm, Nagybörzsöny, Hungary

Lavendulane, 36 mm, San Juan, Chile

Mixite, 70 mm, Cínovec, Czech Republic

Pharmacosiderite

$KFe^{3+}{}_4(AsO_4)_3(OH)_4 . 6 H_2O$

CUBIC ● ● ●

Properties: C – green, yellow-brown, brown; S – white; L – adamantine to greasy; D – transparent to translucent; DE – 2.8; H – 2.5; CL – imperfect; F – uneven; M – cubic crystals, coatings, crusts, granular, massive.
Origin and occurrence: Secondary in granitic pegmatites and in the oxidation zone of ore deposits, where it forms as a product of arsenopyrite and löllingite replacement. Cubic crystals up to 10 mm (3/8 in) across occur in St. Day, Liskeard and Redruth, Cornwall, UK. It also comes from the Majuba Hill mine, Nevada, USA; Horhausen, Germany; and Cap Garrone, France.

Arseniosiderite

$CaFe^{3+}{}_3(AsO_4)_3O_2 . 3 H_2O$

MONOCLINIC ● ●

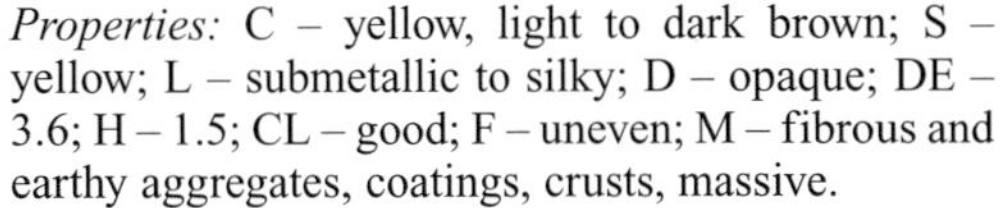

Properties: C – yellow, light to dark brown; S – yellow; L – submetallic to silky; D – opaque; DE – 3.6; H – 1.5; CL – good; F – uneven; M – fibrous and earthy aggregates, coatings, crusts, massive.
Origin and occurrence: Secondary in the oxidation zone of ore deposits and in granitic pegmatites, where it forms as a product of arsenopyrite and löllingite replacement. Rich aggregates occur in Romanché, France; also from the Eureka mine, Tintic, Utah, USA; and Wittichen, Germany.

Lavendulane

$NaCaCu_5(AsO_4)_4Cl . 5 H_2O$

ORTHORHOMBIC ● ●

Properties: C – light blue-purple, blue; S – white; L – vitreous to waxy; D – translucent; DE – 3.5; H – 2.5; CL – good; F – uneven; M – acicular crystals, acicular and earthy aggregates, coatings.
Origin and occurrence: Secondary in the oxidation zone of Co, Cu and Ni deposits, as a result of arsenide weathering, associated with erythrite. Rich acicular aggregates come from Talmessi and Anarak, Iran. Small crystals found in the Blanca mine, San Juan, Chile and in Annaberg, Germany. Crystals up to 4 mm (5/32 in) across occur in the Gold Hill mine, Utah, USA.

Kovdorskite, 15 mm x, Kovdor, Kola, Russia

Kovdorskite

$Mg_5(PO_4)_2(CO_3)(OH)_2 \cdot 4{,}5\ H_2O$

MONOCLINIC ●

Properties: C – light pink-brown, white, blue; S – white; L – vitreous; D – transparent to translucent; DE – 2.6; H – 4; CL – none; F – conchoidal to uneven; M – tabular crystals, massive.
Origin and occurrence: Hydrothermal, associated with magnesite, magnetite and other minerals. Blue and pink-brown crystals up to 25 mm (1 in) across come from the Zheleznyi mine, Kovdor, Kola Peninsula, Russia.

Mixite

$BiCu_6(AsO_4)_3(OH)_6 \cdot 3\ H_2O$

HEXAGONAL ● ●

Properties: C – emerald-green, blue-green, light blue, light green, whitish; S – light green, light blue; L – adamantine to dull; D – translucent; DE – 3.8; H – 3-4; CL – not determined; F – uneven; M – acicular crystals, acicular and earthy aggregates, coatings, massive.
Origin and occurrence: Secondary in the oxidation zone of Bi and Cu deposits, together with bismuthinite. It occurred in Jáchymov, Czech Republic; in Tintic, Utah, USA; Schneeberg and Wittichen, Germany.

Torbernite

$Cu(UO_2)_2(PO_4)_2 \cdot 8\text{-}12\ H_2O$

TETRAGONAL ● ● ●

Properties: C – emerald-green to grass-green; S – light green; L – vitreous to dull, pearly on the cleavage planes; D – transparent to translucent; DE – 3.3; H – 2-2.5; CL – perfect; F – uneven; M – tabular to pyramidal crystals, earthy aggregates, coatings, granular, massive; R – strongly radioactive.
Origin and occurrence: Secondary in the oxidation zone of U deposits, also in pegmatites and sedimentary rocks, resulting from the hydrothermal alteration of uraninite and other U minerals. Emerald-green tabular crystals, several cm across, come from Sabugal, Portugal; Jáchymov, Czech Republic; Shinkolobwe, Zaire; Bois-Noirs, France; Moctezuma, Mexico and many localities in the Colorado Plateau, Utah, USA. Beautiful druses of crystals up to 20 mm ($^{25}/_{32}$ in) across, are found in the Margabal mine, Aveyron, France.
Application: U ore.

Torbernite, 120 mm, Katanga, Zair

Autunite, 15 mm xx, Troncasa, Portugal

Nováčekite, 70 mm, Brumado, Brazil

Autunite

$Ca(UO_2)_2(PO_4)_2 \cdot 10\text{-}12\ H_2O$

TETRAGONAL ● ● ●

Properties: C – light to dark yellow, yellow-green to green; S – light yellow; L – vitreous to dull; D – transparent to translucent; DE – 3.1; H – 2-2.5; CL – perfect; F – uneven; M – tabular crystals, foliated, scaly, earthy and pulverulent aggregates, coatings, massive; LU – yellow-green; R – strongly radioactive.

Origin and occurrence: Secondary in the oxidation zone of U deposits, in pegmatites and in some U-rich sedimentary rocks, as a result of hydrothermal alteration of uraninite and other U minerals. It is frequently associated with torbernite and other U secondary minerals. Tabular crystals up to 30 mm ($1^3/_{16}$ in) across come from Schneeberg and Johanngeorgenstadt, Germany and Autun, France. It is also known from Rum Jungle, Northern Territory, Australia; St. Austel, Cornwall, UK; Mount Spokane, Washington, USA; and Jáchymov, Czech Republic.

Application: U ore.

Uranocircite

$Ba(UO_2)_2(PO_4)_2 \cdot 10\ H_2O$

TETRAGONAL ● ●

Properties: C – light to dark yellow, light yellow-green; S – light yellow; L – vitreous to dull, pearly on the cleavage planes; D – transparent to translu-

Uranocircite, 78 mm, Minas Gerais, Brazil

Zeunerite, 5 mm xx, Cínovec, Czech Republic

cent; DE – 3.5; H – 2-2.5; CL – perfect; F – uneven; M – tabular crystals, foliated and earthy aggregates, pulverulent coatings, massive; LU – green; R – strongly radioactive.
Origin and occurrence: Secondary in the oxidation zone of U deposits. Yellow tabular crystals up to 10 mm ($^3/_8$ in) across occurred in Dametice, Czech Republic; Bergen and Wölsendorf, Germany; and in the Sao Pedro mine, Suaçui, Minas Gerais, Brazil.

Nováčekite

$Mg(UO_2)_2(AsO_4)_2 \cdot 10\ H_2O$

TETRAGONAL ● ●

Properties: C – straw-yellow, light yellow; S – light yellow; L – vitreous to dull; D – transparent to translucent; DE – 3.7; H – 2.5; CL – perfect; F – uneven; M – tabular crystals, lamellar, earthy and pulverulent aggregates, massive; LU – dark green; R – strongly radioactive.
Origin and occurrence: Secondary in the oxidation zone of U deposits. Tabular crystals up to 50 mm (2 in) across come from the Pedra Preta Mine, Brumado, Bahía, Brazil. Lamellar aggregates are known from Zálesí, Czech Republic; Aldama, Chihuahua, Mexico; and Wittichen, Germany.

Zeunerite

$Cu(UO_2)_2(AsO_4)_2 \cdot 10\text{-}16\ H_2O$

TETRAGONAL ● ●

Properties: C – emerald-green, yellow-green; S – light green; L – vitreous to dull; D – transparent to translucent; DE – 3.4; H – 2.5; CL – perfect; F – uneven; M – tabular crystals, foliated aggregates, massive; R – strongly radioactive.
Origin and occurrence: Secondary in the oxidation zone of U deposits. Tabular crystals up to 30 mm ($1^3/_{16}$ in) across, come from the Pedra Preta Mine, Brumado, BahRa, Brazil. It is also known from Zálesí, Czech Republic and Schneeberg, Germany.

Carnotite

$K_2(UO_2)_2(VO_4)_2 \cdot 3\ H_2O$

MONOCLINIC ● ● ●

Properties: C – light to dark yellow, yellow-green; S – light yellow; L – dull; D – transparent to translucent, opaque; DE – 4.9; H – not determined; CL – perfect; F – uneven; M – earthy aggregates, massive; R – strongly radioactive.
Origin and occurrence: Secondary in the oxidation zone of sedimentary U deposits, typically associated

Carnotite, 55 mm, Utah, U.S.A.

with tyuyamunite. Platy crystals, up to 2 mm ($^1/_{16}$ in) across found in the Mashamba West mine, Zaire. Earthy and pulverulent aggregates occur in many localities in the Colorado Plateau, e.g. Paradox Valley, Colorado and La Sal, Utah, USA. Also known from Tyuya Muyun, Uzbekistan and Radium Hill, Southern Australia.
Application: U and V ore.

Tyuyamunite

$Ca(UO_2)_2(VO_4)_2 \cdot 5\text{-}8\ H_2O$

ORTHORHOMBIC ● ● ●

Properties: C – yellow-green, canary yellow; S – light yellow; L – silky to adamantine; D – translucent to opaque; DE – 3.6; H – 2; CL – perfect; F – uneven; M – earthy aggregates, massive; LU – weak yellow-green; R – strongly radioactive.
Origin and occurrence: Secondary in the oxidation zone of sedimentary U deposits, together with carnotite. Common earthy and pulverulent aggregates occur in many localities in the Colorado Plateau, e.g. Paradox Valley, Colorado and Red Creek, Utah, USA. It was described from Tyuya Muyun, Uzbekistan.
Application: U and V ore.

Tyuyamunite, 80 mm, Fergana, Uzbekistan

9. Silicates

Phenakite

Be_2SiO_4

TRIGONAL ● ● ●

Properties: C – colorless, white, yellowish; S – white; L – strong vitreous; D – transparent to translucent; DE – 3.0; H – 8; CL – imperfect; F – conchoidal; M – long prismatic to tabular crystals, radial aggregates, granular.
Origin and occurrence: Magmatic in granitic pegmatites; hydrothermal in greisens; metamorphic in mica schists, associated with beryl, chrysoberyl, apatite and quartz. Prismatic crystals up to 250 mm ($9^{13}/_{16}$ in) long occurred in Kragerö, Norway. It is also known from Sao Miguel de Piracicaba, Minas Gerais, Brazil in crystals, up to 100 mm (4 in) long. The other localities are Habachtal, Austria; Malyshevo, Russia; and Anjanabonoina, Madagascar.
Application: sporadically cut as a gemstone.

Phenakite, 200 mmx, Kragerö, Norway

Willemite

Zn_2SiO_4

TRIGONAL ● ● ●

Properties: C – white, yellowish, gray, green; S – white; L – vitreous; D – translucent; DE – 4.0; H – 5.5; CL – good; F B conchoidal to uneven; M – prismatic to tabular crystals, radial aggregates, granular; LU – distinct light green.
Origin and occurrence: Metamorphic in marbles; secondary in the oxidation zone of ore deposits, associated with zincite, franklinite, hemimorphite and smithsonite. Crystals up to 100 mm (4 in) across come from Franklin, New Jersey, USA; Mont St.-Hilaire, Quebec, Canada.
Application: as Zn ore.

Topaz, 60 mm, Thomas Range, U.S.A.
Willemite, 76 mm, Tsumeb, Namibia

Forsterite, 40 mm, Suppat, Pakistan

Forsterite

OLIVINE GROUP

Mg_2SiO_4

ORTHORHOMBIC ● ● ● ●

Properties: C – yellowish, greenish, colorless; S – white; L – vitreous; D – transparent to opaque; DE – 3.3; H – 6.5-7; CL – good; F – conchoidal to uneven; M – tabular to prismatic crystals, granular.

Origin and occurrence: Metamorphic in regionally and contact metamorphosed dolomites. Typical rock-forming mineral, associated with enstatite, spinel, phlogopite and chlorite. Green gemmy crystals up to 80 mm (3$^1/_8$ in) long come from Suppat, Pakistan. It also occurred in Crestmore, California, USA; Mount Timobly, British Columbia, Canada; and Monte Somma, Italy.

Olivine

OLIVINE GROUP

$(Mg,Fe)_2SiO_4$

ORTHORHOMBIC ● ● ● ● ●

Varieties: chrysolite

Properties: C – green, yellow-green (chrysolite), brown-green to black-green; S – white; L – vitreous; D – translucent to opaque; DE – 3.3-3.6; H – 6.5-7; CL – good; F B conchoidal to uneven; M – imperfect crystals, granular.

Origin and occurrence: Magmatic in some ultrabasic rocks, e.g. dunites, lherzolites, peridotites, gabbros and in meteorites. Typical rock-forming mineral, usually associated with diopside, magnetite and pyrope. Classic locality of chrysolite is Zebirget Island in the Red Sea, Egypt, where tabular crystals

Olivine, 60 mm, Smrčí, Czech Republic

up to 100 mm (4 in) across occurred. It is also known from the San Carlos Indian Reservation, Arizona, USA. Olivine was found in many basaltic rocks in Laacher See, Germany; Rockport, Massachusetts, USA and elsewhere.
Application: chrysolite is cut as a gemstone.

Fayalite

OLIVINE GROUP

Fe_2SiO_4

ORTHORHOMBIC ● ● ●

Properties: C – black-green to black; S – gray; L – dull to vitreous; D – opaque; DE – 4.2; H – 6.5-7; CL – good; F B conchoidal to uneven; M – imperfect prismatic crystals, granular.
Origin and occurrence: Magmatic in granitic pegmatites, granites and syenites, associated with orthoclase, gadolinite-(Y) and epidote; rare metamorphic. Poorly developed crystals up to 150 mm (6 in) long found in pegmatites near Baveno, the Alps, Italy. It is also known from Strzegom, Poland and the Sawtooth Batholith, Idaho, USA.

Fayalite, 80 mm, Rockport, U.S.A.

Tephroite

OLIVINE GROUP

Mn_2SiO_4

ORTHORHOMBIC ● ●

Properties: C – gray, olive-green, red-brown; S – white; L – dull to vitreous; D – translucent to transparent; DE – 4.2; H – 6; CL – good; F B conchoidal to uneven; M – prismatic crystals, granular.
Origin and occurrence: Metamorphic in skarns and Mn-rich metamorphosed sediments, together with rhodonite, franklinite and spessartine. Granular aggregates and perfect crystals up to 50 mm (2 in) across known from Franklin, New Jersey, USA and Langban, Sweden. It comes also from Tarnobrzeg, Poland, in crystals, up to 80 mm ($3^1/_8$ in) across.

Tephroite, 60 mm, Harstigen, Sweden

Pyrope, 3 mm grains, Třebívlice, Czech Republic

Pyrope

GARNET GROUP

$Mg_3Al_2Si_3O_{12}$

CUBIC ● ● ● ●

Properties: C – red to purple-red, light purple, black-brown; S – white; L – vitreous; D – transparent to translucent; DE – 3.5; H – 7-7.5; CL – none; F B conchoidal to uneven; M – isometric crystals, granular.

Origin and occurrence: Magmatic in some ultrabasic rocks, e.g. lherzolites, peridotites, kimberlites, eclogites and serpentinites; metamorphic in quartzites; also known from placers. It is associated with diopside, magnetite and diamond. It comes from many localities in ultrabasic rocks, like Třebenice and Mirunice, Czech Republic; Zöblitz, Germany; Madras, India; and Kimberley, South Africa. Crystals up to 250 mm ($9^{13}/_{16}$ in) across were found in Dora Maria, the Alps, Italy.

Application: cut as a gemstone.

Almandine

GARNET GROUP

$Fe_3Al_2Si_3O_{12}$

CUBIC ● ● ● ●

Properties: C – red to purple-red, black-brown; S – white; L – vitreous; D – transparent to translucent; DE – 4.3; H – 7; CL – none; F B conchoidal to uneven; M B well-formed crystals, granular.

Origin and occurrence: Metamorphic in regionally metamorphosed rocks, as chlorite schists, gneisses,

Almandine, 38 mm x, Ötztal, Austria

Spessartine, 96 mm, Gilgit, Pakistan

Grossular, 17 mm x, Asbestos, Canada

mica schists and migmatites; magmatic in some granites and pegmatites; also in placers. Well-developed crystals up to 150 mm (6 in) across are known from Ishikawa pegmatites, Japan and Shingus, Pakistan. It comes from many mica schists and gneisses as crystals, up to about 50 mm (2 in), like Fort Wrangel, Alaska, USA; Ötztal, Austria; and Bodö, Norway. It occurs in placers near Ratnapura, Sri Lanka.
Application: cut as a gemstone, abrasive material.

Spessartine

GARNET GROUP
$Mn_3Al_2Si_3O_{12}$

CUBIC ● ● ● ●

Properties: C – red, orange, light brown to yellowish; S – white; L – vitreous; D – transparent to translucent; DE – 4.3; H – 7-7.5; CL – none; F B conchoidal to uneven; M – perfect crystals, granular.
Origin and occurrence: Magmatic in granitic pegmatites and some granites; hydrothermal in cavities in rhyolites; metamorphic in some skarns and Mn-rich metamorphic rocks. Perfect crystals up to 30 mm ($1^3/_{16}$ in) across were found in granitic pegmatites in the Hercules mine, Ramona, California, USA; near Marienfluss river, Namibia; and in rhyolite cavities in Nathrop, Colorado, USA. Gemmy crystal fragments up to 50 mm (2 in) across were recently found in an undisclosed locality in Minas Gerais, Brazil.
Application: cut as a gemstone.

Grossular

GARNET GROUP
$Ca_3Al_2Si_3O_{12}$

CUBIC ● ● ● ●

Varieties: hessonite, tsavorite

Properties: C – red, green (tsavorite), orange, red-brown (hessonite) to colorless; S – white; L – vitreous; D – transparent to tranlucent; DE – 3.4; H – 6.5-7; CL – none; F B conchoidal to uneven; M – perfect crystals, granular.
Origin and occurrence: Metamorphic in Ca-rich, contact metamorphic rocks, skarns, rodingites; hydrothermal along the cracks in these rocks, associated with diopside, vesuvianite, wollastonite, scapolite and epidote. Perfect crystals about 30 mm ($1^3/_{16}$ in) across occur in the Jeffrey quarry, Asbestos, Quebec, Canada and Sierra de las Cruces, Coahuila, Mexico. Tsavorite crystals up to 50 mm (2 in) across come from the Tsavo National Park, Kenya and Merelani Hills, Arusha, Tanzania. The other well-known localities are Ala, Italy and Ciclova, Romania. Grossular crystals up to 100 mm (4 in) in size in Vápenná, Czech Republic; Xalostoc, Mexico; and Sandaré, Mali.
Application: cut as a gemstone.

Andradite, 30 mm xx, Graham Co., U.S.A.

Andradite

GARNET GROUP

$Ca_3Fe^{3+}{}_2Si_3O_{12}$

CUBIC ● ● ● ●

Varieties: demantoid, melanite

Properties: C – dark red, black-brown, brown, green to yellow-green (demantoid), black-brown to black (melanite); S – white; L – vitreous; D – transparent to translucent; DE – 3.9; H – 6.5-7; CL – none; F B conchoidal to uneven; M B well-developed crystals, granular.

Origin and occurrence: Metamorphic in Ca and Fe rich contact metamorphosed rocks, skarns, rodingites; hydrothermal along the cracks of these rocks; magmatic in some alkaline igneous rocks, usually in the same localities as grossular. Crystals up to 40 mm ($1^{9}/_{16}$ in) across come from Sinerechenskoye, Russia. Fine crystals were also found in the Namgar mine, Usakos, Namibia and Ciclova, Romania. Fine melanite crystals occur in Magnet Cove, Arkansas, USA. Demantoid crystals up to 30 mm ($1^{3}/_{16}$ in) across known from Val Malenco, Italy and in the Bobrovka river basin, Ural mountains, Russia.

Application: demantoid is cut as a gemstone.

Demantoid, 3 mm xx, Bobrovka, Russia

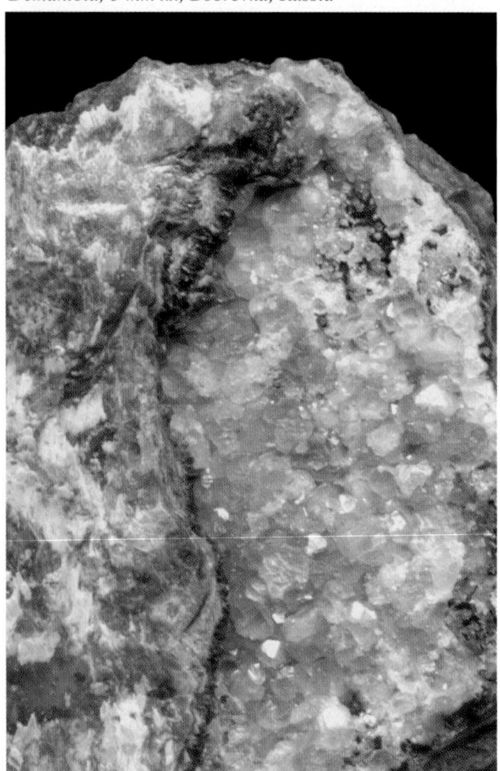

Melanite, 5 mm xx, Rudnyi, Kazakhstan

Uvarovite, 2 mm xx, Sarany, Russia

Uvarovite

GARNET GROUP

$Ca_3Cr_2Si_3O_{12}$

CUBIC ● ●

Properties: C – dark emerald-green; S – white; L – vitreous; D – transparent to translucent; DE – 3.9; H – 6.5-7; CL – none; F B conchoidal to uneven; M – perfect crystals, granular.

Origin and occurrence: Metamorphic and also hydrothermal are almost only limited to rocks with increased Cr content, ultrabasic rocks with chromite, serpentinites and skarns. It occurs as crystals up to 8 mm (5/16 in) across along cracks in chromite in Sarany, Ural mountains, Russia. Crystals up to 20 mm (25/32 in) across, come from Outokumpu, Finland. It is also known from Orford, Quebec, Canada.

Zircon

$ZrSiO_4$

TETRAGONAL ● ● ●

Varieties: jargon, hyacinth.

Properties: C – yellow (jargon), brown, yellow-brown, red-orange (hyacinth), red to colorless; S – white; L – vitreous, greasy to adamantine; D – transparent to translucent; DE – 4.7; H – 7.5; CL – imperfect; F B conchoidal to uneven; M – perfect long and short prismatic crystals, granular; LU – yellow; R B sometimes radioactive and usually metamict.

Origin and occurrence: Magmatic and metamorphic as an accessory mineral in different rock types; rarely hydrothermal in the Alpine-type and quartz veins; also in sediments and placers. Prismatic crystals up to 30 cm (12 in) across, occur in syenite pegmatites in Renfrew and Bancroft, Ontario, Canada. It is also known from Miass, Ural mountains and from Mount Vavnbed, Lovozero massif, Kola Peninsula, Russia. Classic occurrences in granitic pegmatites are Alto Ligonha, Mozambique; Arendal, Norway; Ytterby, Sweden and Jaguaraçu, Minas Gerais, Brazil. Gemmy zircons of different colors up to 80 mm (3 1/8 in) in size come from placers near Ratnapura, Sri Lanka and elsewhere.

Application: hyacinth and jargon are cut as gemstones, Zr ore.

Zircon, 20 mm x, Vishnevogorsk, Ural Mts., Russia

Eulytine, 2 mm xx, Schneeberg, Germany

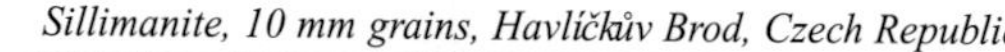

Sillimanite, 10 mm grains, Havlíčkův Brod, Czech Republic

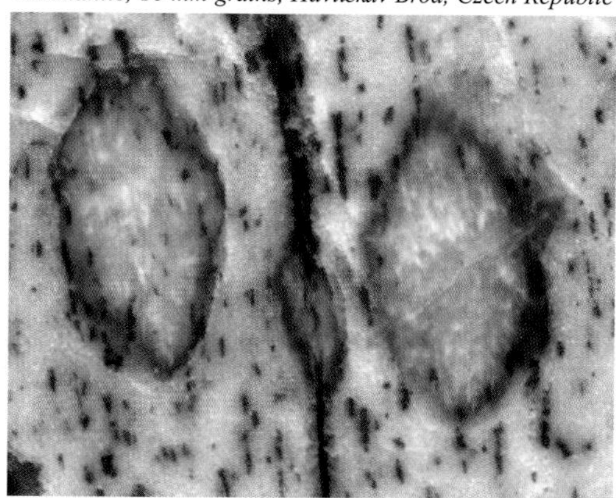

Eulytine

$Bi_4Si_3O_{12}$

CUBIC ● ●

Properties: C – brown, yellow, gray, colorless; S – white; L – greasy; D – translucent; DE – 6.6; H – 4.5; CL – good; F B conchoidal to uneven; M – small dipyramidal crystals, radial aggregates, granular.
Origin and occurrence: Secondary in Bi deposits, typically associated with bismuth. Its small crystals come from Horní Slavkov, Czech Republic; Schneeberg and Johanngeorgenstadt, Germany.

Euclase

$BeAlSiO_4(OH)$

MONOCLINIC ● ●

Properties: C – colorless, white, greenish, blue; S – white; L – vitreous; D – transparent to translucent; DE – 3.1; H – 7.5; CL – good; F B conchoidal to

Euclase, 42 mm, Minas Novas, Brazil

uneven; M – long prismatic crystals, radial aggregates, granular.
Origin and occurrence: Hydrothermal occurs in pegmatites, greisens and in quartz and Alpine-type veins; rare in placers. Well-formed crystals up to 80 mm (3$^{1}/_{8}$ in) in size known from Santa do Encoberto, Minas Gerais, Brazil. Blue crystals up to 50 mm (2 in) across occurred in the Last Hope Mine, Karoi, Zimbabwe. It also come from the sediments of Sanarka River, Ural mountains, Russia. Dark blue crystals up to 150 mm (6 in) across were found recently in the Chivor Mine, Colombia.
Application: locally cut as a gemstone.

Sillimanite

Al_2SiO_5

ORTHORHOMBIC ● ● ● ● ●

Properties: C – white, gray, greenish, yellowish; S – white; L – vitreous to dull; D – transparent to translucent; DE – 3.3; H – 6.5-7.5; CL – good; F – uneven; M – long prismatic crystals, fibrous aggregates, granular.
Origin and occurrence: Almost exclusively metamorphic in gneisses and migmatites; only rare magmatic in pegmatites and granites; also in placers. Typical rock-forming mineral, very commonly associated with andalusite. Typical localities are Bodenmais, Germany and Marsíkov, Czech Republic. Gemmy crystals up to 20 mm ($^{25}/_{32}$ in) long are known from Rakwana-Deniyaya, Sri Lanka.

Andalusite

Al_2SiO_5

ORTHORHOMBIC ● ● ● ●

Varieties: chiastolite, viridine

Properties: C – pink, red-brown, red, gray, whitish, green (viridine); S- white; L – vitreous to dull; D – translucent to transparent; DE – 3.2; H – 6.5-7.5; CL – good; F – uneven; M – prismatic crystals, fibrous aggregates, granular, massive.
Origin and occurrence: Metamorphic in regionally and contact metamorphosed rocks; magmatic in pegmatites and granites, hydrothermal in quartzites. Typical rock-forming mineral, commonly associated with sillimanite, corundum and cordierite. Renowned localities are Lisens, the Alps, Austria, and Bimbowrie, South Australia. Green gemmy crystals come from Morro do Chapeú, Bahía, Brazil. Viridine is known from Darmstadt, Germany.

Andalusite, 28 mm x, Aracuay, Minas Gerais, Brazil

Chiastolith, 20 mm, Bimbowrie, Australia

Kyanite

Al_2SiO_5

TRICLINIC ● ● ● ●

Properties: C – blue, gray, white, green, dark gray, colorless; S – white; L – vitreous to dull; D – transparent to translucent; DE – 3.6; H – 4.5-7.5; CL – good; F – uneven; M – prismatic to tabular crystals, fibrous aggregates, granular, massive.

Origin and occurrence: Almost only metamorphic in regionally metamorphosed rocks, mica schists, gneisses, granulites and eclogites; less frequently magmatic in pegmatites and granites; rarely hydrothermal in quartz veins. Typical rock-forming mineral, associated with andalusite and sillimanite. Blue columnar aggregates and crystals up to 150 mm (6 in) long occur in Barra do Salinas, Minas Gerais, Brazil. Other renowned localities are Pizzo Forno, Switzerland; Prilep, Macedonia; and Keivy, Kola Peninsula, Russia.

Kyanite, 87 mm, Minas Gerais, Brazil

Topaz, 18 mm xx, Ghundao Hill, Pakistan

Topaz

$Al_2SiO_4(F,OH)$

ORTHORHOMBIC ● ● ●

Varieties: pycnite

Properties: C – colorless, blue, yellow, gray, white, greenish, pinkish, red; S – white; L – vitreous; D – transparent to translucent; DE – 3.6; H- 8; CL – good; F – uneven; M – perfect prismatic to tabular crystals, radial and columnar aggregates, granular.

Origin and occurrence: Magmatic in pegmatites and granites; hydrothermal in greisens, in rhyolite cavities, in quartz veins, also in placers. Topaz crystals in pegmatites are occasionally very large, like the crystal measuring 80 x 60 x 60 cm (31½ x 24 x 24 in) across from Fazenda do Funil; crystals

Topaz, 49 mm, Ramona, U.S.A.

up to 30 cm (12 in) from Virgem da Lapa, both Minas Gerais, Brazil. Blue, brownish and bicolored crystals up to 40 cm (15¾ in) long come from Volodarsk Volynskii, Ukraine. Other topaz localities are Iveland, Norway; Murzinka, Ural mountains, Russia; Little Three mine, Ramona, California and Pikes Peak, Colorado, USA; and Gilgit, Pakistan; Spitzkopje, Namibia. Orange to red topaz (imperial topaz) comes from quartz veins near Ouro Preto, Minas Gerais, Brazil. Pink topaz crystals up to 70 mm (2¾ in) found at Mount Ghundao, Mardan, Pakistan. Important topaz specimens were also found in Schneckenstein, Germany; Thomas Range, Utah, USA; and Nerchinsk, Siberia, Russia. Columnar aggregates of pycnite come from Cínovec, Czech Republic and Altenberg, Germany.
Application: cut as a gemstone.

Pycnite, 60 mm, Cínovec, Czech Republic

Staurolite, 80 mm, Keivy, Kola, Russia

Staurolite

$Fe_2Al_9Si_4O_{22}(OH)_2$

ORTHORHOMBIC ● ● ● ●

Properties: C – dark to light brown, yellow; S – white; L – dull to vitreous; D – transparent to almost opaque; DE – 3.7; H – 7-7.5; CL – good; F – uneven; M – prismatic to tabular crystals and their combinations, granular.

Origin and occurrence: Metamorphic in regionally metamorphosed rocks, gneisses and mica schists; rare magmatic in granites; also in placers, commonly associated with almandine, andalusite and kyanite. Crystals up to 50 mm (2 in) across come from Pizzo Forno, Switzerland. It is also known from Rio Arriba, New Mexico, USA. Its cross-like twins, up to 200 mm ($7\frac{7}{8}$ in) across, occur in Keivy, Kola Peninsula, Russia and Morbihan, France.

Sapphirine, 26 mm, Vohimena, Madagascar

Sapphirine

$Mg_2Al_4SiO_{10}$

MONOCLINIC ● ●

Properties: C – dark to light blue, green; S – white; L – vitreous to dull; D – transparent to translucent; DE – 3.5; H – 7.5; CL – imperfect; F – uneven; M – tabular crystals, granular.

Origin and occurrence: Metamorphic in regionally metamorphosed rocks, rich in Al and Mg and poor in Si, associated with spinel and corundum. Crystals up to 40 mm ($1\frac{9}{16}$ in) across come from Fiskenaesset, Greenland; Betroka and Androy, Madagascar. Other localities include Val Codera, Italy and Enderby Land, Antarctica.

Chondrodite, 15 mm x, Brewster, U.S.A.

Clinohumite, 19 mm, Kukh-i-lal, Tadzhikistan

Chondrodite

$Mg_5Si_2O_8(OH,F)_2$

MONOCLINIC ● ● ●

Properties: C – yellow, greenish, brown; S – white; L – vitreous to dull; D – transparent to translucent; DE – 3.2; H – 6-6.5; CL – imperfect; F – uneven; M – crystals of different habits, granular.
Origin and occurrence: Metamorphic in contact and regionally metamorphosed carbonate rocks; rare magmatic in carbonatites, typically associated with spinel, chlorite and phlogopite. Perfect crystals up to 50 mm (2 in) long come from the Tilly Foster mine, Brewster, New York, USA. Other localities are Pargas, Finland; Monte Somma, Italy; and Riverside, California, USA.

Clinohumite

$Mg_9Si_4O_{16}(OH,F)_2$

MONOCLINIC ● ● ●

Properties: C – yellow, red, brown, white; S – white; L – vitreous to dull; D – transparent to translucent; DE – 3.3; H – 6; CL – imperfect; F – uneven; M – crystals of different habits, granular.
Origin and occurrence: Metamorphic in contact and regionally metamorphosed carbonate rocks, serpentinites and talc schists, associated with spinel, chlorite, forsterite, serpentine and phlogopite. Gemmy yellow crystals up to 30 mm (1$^{3}/_{16}$ in) across come from Kukh-i-Lal, Pamir, Tajikistan. Other localities are Pargas, Finland; Monte Somma, Italy; and Jensen quarry, California, USA.

Chapmanite

$SbFe^{3+}{}_2Si_2O_8(OH)$

MONOCLINIC ● ●

Properties: C – olive-green to dark yellow; S – yellowish to yellow; L -dull; D – translucent; DE – 3.7; H – 2.5; CL – imperfect; F – uneven; M – elongated crystals, massive.
Origin and occurrence: Hydrothermal and secondary in the cracks of rocks, sometimes with stibnite. Its massive aggregates occur in Smilkov near Votice, Czech Republic and in the Keely mine, Cobalt, Ontario, Canada.

Chapmanite, 4 mm, Pernek, Slovakia

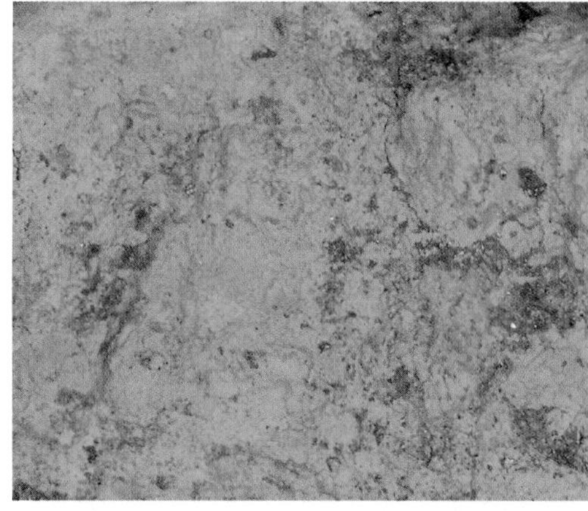

Braunite, 60 mm, Langban, Sweden

Braunite

$Mn^{2+}Mn^{3+}{}_6SiO_{12}$

TETRAGONAL ● ● ●

Properties: C – black, black-gray, black-brown; S – gray; L – submetallic to metallic; D – opaque; DE – 4.7; H – 6-6.5; CL – perfect; F – uneven to conchoidal; M – dipyramidal crystals, granular.
Origin and occurrence: Metamorphic in regionally and contact metamorphosed rocks; hydrothermal in sedimentary rocks rich in Mn and in hydrothermal veins, associated with hausmanite, pyrolusite and other Mn minerals. Its perfect crystals up to 70 mm (2¾ in) long, come from Kacharhavee and Tirodi, India. It is also known from Langban, Sweden; Ilfeld and Ilmenau, Germany; St. Marcel, Italy; and Tizi Bashkun, Morocco.
Application: Mn ore.

Thaumasite

$Ca_6Si_2(CO_3)_2(SO_4)_2(OH)_{12} \cdot 24\ H_2O$

HEXAGONAL ● ●

Properties: C – white to colorless; S – white; L – vitreous to dull; D – transparent to translucent; DE – 1.9; H – 3.5; CL – imperfect; F – uneven; M B acicular aggregates, granular, massive; LU – white.
Origin and occurrence: Hydrothermal or metamorphic in contact metamorphosed carbonate rocks, usually associated with other Ca silicates and carbonates, like ettringite and prehnite. Typical localities are Crestmore, California and West Paterson, New Jersey, USA; Langban, Sweden; and N'Chwaning mine No. 2. Kuruman, South Africa.

Titanite

$CaTiOSiO_4$

MONOCLINIC ● ● ● ●

Properties: C – colorless, yellow, brown, green, gray to black; S – white; L – vitreous to dull; D – transparent, translucent to opaque; DE – 3.5; H – 5-5.5; CL – good; F – uneven to conchoidal; M – tabular crystals and their combinations, granular, massive.
Origin and occurrence: Metamorphic and magmatic as a common accessory mineral in many igneous and

Thaumasite, 3 mm aggregate, Maglovec, Slovakia

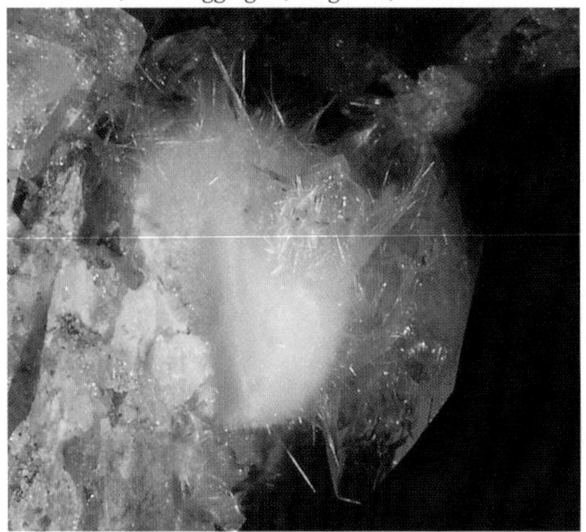

Fersmanite, 10 mm x, Khibiny Massif, Kola, Russia

Titanite, 20 mm xx, Stubachtal, Austria

metamorphic rocks and pegmatites; hydrothermal in the Alpine-type veins; also in placers. The most beautiful crystals up to 180 mm ($7\frac{1}{16}$ in) long occur in Alpine-type veins in Tavetsch and Binntal, Switzerland; Zillertal and Felbertal, Austria; and Dodo, Polar Ural, Russia. Large, poorly developed crystals weighing up to 40 kg (88 lb), come from Eganville, Ontario, Canada and Rossie, New York, USA.

Chloritoid, 30 mm, Ille de Croix, France

Fersmanite

$(Na,Ca)_4(Ti,Nb)_2Si_2O_{11}(OH,F)_2$

TRICLINIC ● ●

Properties: C – light to dark brown; S -white; L – vitreous to dull; D – translucent; DE – 3.5; H – 5-5.5; CL – none; F – uneven to conchoidal; M – tabular crystals.
Origin and occurrence: Magmatic to hydrothermal in alkaline pegmatites together with pectolite, apatite and sulfides. Crystals up to 30 mm ($1\frac{3}{16}$ in) across known from Mount Eveslogchorr, Khibiny massif, Kola Peninsula, Russia.

Chloritoid

$Fe_2Al_4Si_2O_{10}(OH)_2$

MONOCLINIC, TRICLINIC 

Properties: C – dark gray, gray-green to black-green; S – gray; L – vitreous to dull; D – translucent; DE – 3.6; H – 6.5; CL – perfect; F – uneven to conchoidal; M – tabular crystals, foliated and scaly aggregates, granular.
Origin and occurrence: Metamorphic in some mica schists and phyllites; hydrothermal alteration product in lavas. Typical localities are Zermatt, Switzerland; Ottrez, Belgium; and Pregraten, Austria.

Datolite, 30 mm xx, Dalnegorsk, Russia

Datolite
$CaBSiO_4(OH)$

MONOCLINIC ● ● ●

Properties: C – white to colorless, yellowish, greenish, gray; S – white; L – vitreous to dull; D – transparent to translucent; DE – 3.1; H – 5-5.5; CL – none; F – uneven to conchoidal; M – prismatic to tabular crystals, nodules, granular.
Origin and occurrence: Metamorphic and hydrothermal in contact metamorphosed rocks, in cavities in volcanic rocks, in ore veins and pegmatites. It is commonly associated with zeolites, prehnite, calcite and also with tourmaline. Prismatic and tabular crystals up to 100 mm (4 in) long occur in Dalnegorsk, Russia. Other typical localities are West Paterson, New Jersey; Keweenaw Peninsula, Michigan, USA; Haslach, Germany; and Charcas, San Luis Potosi, Mexico.

Gadolinite-(Y)
$Y_2FeBe_2Si_2O_{10}$

MONOCLINIC ● ●

Properties: C – black, dark red, brown, greenish; S – gray-green; L – vitreous to greasy, D – transparent to translucent; DE – 4.4; H – 6.5-7; CL – none; F –

Gadolinite-(Y), 60 mm, Hitterö, Norway

Howlite, 30 mm, California, U.S.A.

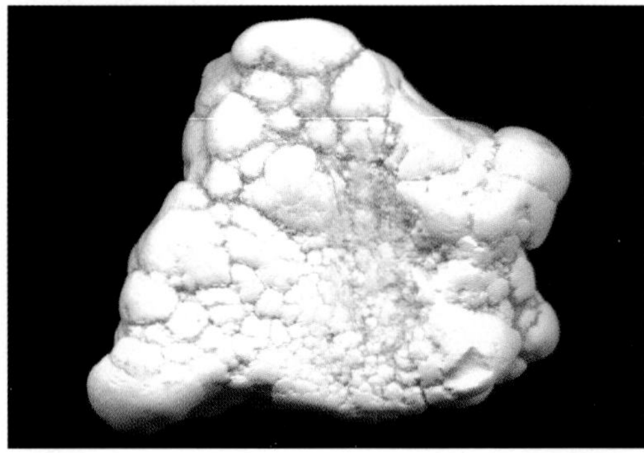

Dumortierite, 60 mm, Dehesa, U.S.A.

conchoidal; M – prismatic crystals, granular; R – locally metamict.
Origin and occurrence: Magmatic in granites and pegmatites, locally associated with fluorite and allanite; rare hydrothermal in the Alpine-type veins. Aggregates, up to 40 cm (15¾ in) across come from Hitterö, Norway. Barringer Hill and Clear Creek, Texas, USA yielded aggregates, weighing up to 90 kg (198 lb). Other famous localities are Blatchford Lake, Northwest Territories, Canada; Baveno, the Alps, Italy; and Iveland, Evje, Norway.

Kornerupine

$Mg_4Al_6(Si,Al,B)_5O_{21}(OH)$

ORTHORHOMBIC ● ●

Properties: C – colorless, white, greenish, gray, brown; S – white; L – vitreous to dull; D – transparent to translucent; DE – 3.3; H – 6.5-7; CL – none; F – conchoidal; M – prismatic crystals, radial aggregates, granular.
Origin and occurrence: Metamorphic in strongly metamorphosed rocks, as granulites. Crystals up to 230 mm (9¹/₁₆ in) across known from Fiskenaesset, Greensland. It also occurs in Lac Ste-Marie, Quebec, Canada; Waldheim, Germany; and Itrongay, Madagascar.

Dumortierite

$(Al,Ti,Mg)Al_6BSi_3O_{16}(O,OH)_2$

ORTHORHOMBIC ● ● ●

Properties: C – purple, pink, blue, brown; S – white; L – vitreous to dull; D – transparent to translucent; DE – 3.4; H – 8.5; CL – good; F – uneven; M – prismatic crystals, radial and acicular aggregates, massive.
Origin and occurrence: Metamorphic in some Al-rich metamorphic rocks, e.g. in migmatites and gneisses; magmatic in granites and pegmatites; hydrothermal in altered rocks. It comes from Dehesa, California; Rochester, Nevada, USA; Beaunan, France; crystals are known from the vicinity of Kutná Hora, Czech Republic and Soavina, Madagascar.

Howlite

$Ca_2B_5SiO_9(OH)_5$

MONOCLINIC ● ●

Properties: C – white; S – white; L – vitreous to dull; D – transparent to translucent; DE – 2.5; H – 3.5; CL – good; F – uneven; M – tabular crystals, nodules, massive.
Origin and occurrence: Hydrothermal in borate deposits, associated with ulexite and colemanite. Crystals, several cm across, come from the vicinity of Bras d'Or Lake, Nova Scotia, Canada. It occurs also in Lang and Daggett, California, USA.

Kornerupine, 90 mm, Lac Ste.-Marie, Canada

Cuprosklodowskite, 82 mm, Musonoi, Zair

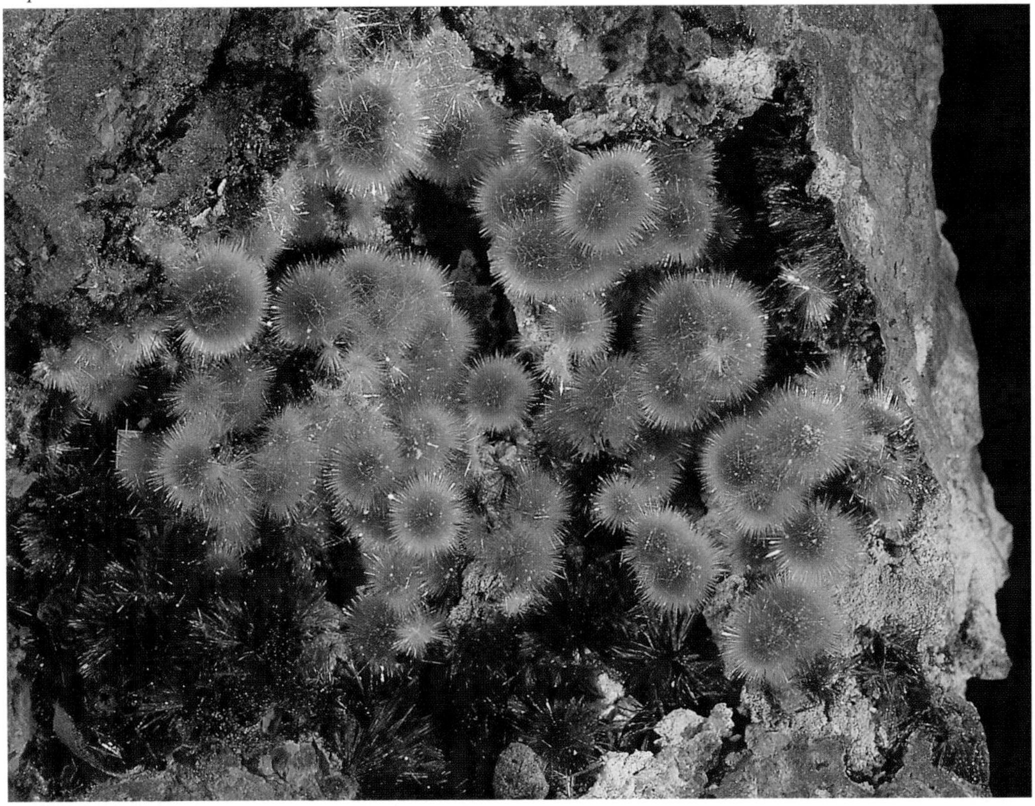

Cuprosklodowskite

$Cu(UO_2)_2Si_2O_7 . 6 H_2O$

MONOCLINIC ● ●

Properties: C – various hues of green; S – greenish; L – vitreous to dull; D – transparent to translucent; DE – 3.8; H – 4; CL – good; F – uneven; M – radial, acicular aggregates, thin coatings, granular, massive; R – strong radioactive.

Origin and occurrence: Secondary in the oxidation zone of U deposits, associated with autunite, torbernite, uranophane and other U secondary minerals. Crystals, up to several cm long, come from

Uranophane, 125 mm, Brewster, Texas, U.S.A.

Kasolite, 30 mm, Musonoi, Zair

the Mashamba West mine, Musonoi, Zaire. It also occurred in Jáchymov, Czech Republic.

Uranophane

$Ca(UO_2)_2Si_2O_7 . 6\,H_2O$

MONOCLINIC ● ● ●

Properties: C – various hues of yellow to brown; S – yellowish; L – vitreous to dull; D – transparent to translucent; DE – 3.9; H – 2.5; CL – good; F – uneven; M – radial, acicular aggregates, thin coatings, granular, massive; LU – yellow-green; R – strong radioactive.
Origin and occurrence: Secondary in the oxidation zone of U deposits as a product of uraninite alteration, associated with autunite, torbernite and other secondary U minerals. Needles up to 10 mm (3/8 in) long, come from Musonoi, Shinkolobwe, Zaire. It is also known from the Faraday mine, Bancroft, Ontario, Canada. It occurs in Wölsendorf, Germany, as well as in Jáchymov, Czech Republic.

Kasolite

$Pb(UO_2)SiO_4 . H_2O$

MONOCLINIC ● ●

Properties: C – various hues of yellow, green to brown; S – yellowish; L – vitreous to dull; D – translucent to opaque; DE – 6.2; H – 4.5; CL – good; F – uneven; M – prismatic crystals, radial and acicular aggregates, thin coatings, granular, massive; R – strong radioactive.
Origin and occurrence: Secondary mineral in the oxidation zone of U deposits. It is a product of uraninite alteration and is associated with uranophane, torbernite and U hydroxides. Crystals up to 10 mm (3/8 in) long occur in Shinkolobwe, Kasolo, Zaire and Mounana, Gabon.

Akermanite

$Ca_2MgSi_2O_7$

TETRAGONAL ● ● ●

Properties: C – white, gray, green, brown; S – white; L – vitreous to dull; D – transparent to translucent; DE – 2.9; H – 5.5; CL – good; F – uneven to conchoidal; M – short prismatic crystals, granular.
Origin and occurrence: Magmatic in volcanic basic rocks; metamorphic in contact metamorphosed marbles. A typical representative of localities in marbles is Crestmore, California, USA. It occurs also in Ca-rich volcanic rocks in Velardena, Mexico.

Akermanite, 5 mm xx, Mt. Vesuvius, Italy

Gehlenite

$Ca_2Al_2SiO_7$

TETRAGONAL ● ● ●

Properties: C – white, gray, yellowish; S – white; L – vitreous to dull; D – transparent to translucent; DE – 3.0; H – 5-6; CL – good; F – uneven to conchoidal; M – short prismatic crystals, granular.
Origin and occurrence: Magmatic in volcanic basic rocks; metamorphic in contact metamorphic marbles. It occurs in Crestmore, California, USA; Monzoni, Italy; Oraviza, Romania and elsewhere.

Gehlenite, 80 mm, Vata de Sus, Romania

Ilvaite, 20 mm xx, Dalnegorsk, Russia

Bertrandite, 36 mm, Mt.Antero, U.S.A.

Ilvaite

$CaFe^{2+}{}_2Fe^{3+}Si_2O_8(OH)$

ORTHORHOMBIC ● ● ●

Properties: C – black to black-gray; S – black; L – submetallic to dull; D – opaque; DE – 4.1; H – 5.5-6; CL – good; F – uneven; M – prismatic crystals, granular.
Origin and occurrence: Metamorphic and hydrothermal in contact metamorphosed Fe, Zn and Cu deposits. Prismatic crystals up to 100 mm (4 in) long come from Rio Marina, Elba, Italy. Crystals up to 30 cm (12 in) long are known from Serifos Island, Greece. Crystals, several cm long occur also in Dalnegorsk, Russia and in the Laxey mine, Idaho, USA.

Bertrandite

$Be_4Si_2O_7(OH)_2$

ORTHORHOMBIC ● ● ●

Properties: C – colorless, white, yellowish; S – white; L – vitreous to dull; D – transparent to translucent; DE – 2.6; H – 6-7; CL – perfect; F – uneven; M – tabular crystals and their combinations, radial aggregates, granular, massive.
Origin and occurrence: Hydrothermal in granitic pegmatites, greisens and in hydrothermal veins, together with beryl, also as pseudo-morphs after beryl. Tabular crystals up to 50 mm (2 in) across occur in Conselheira Pena, Minas Gerais, Brazil. It is known from Kounrad and Kara-Oba, Kazakhstan in crystals up to 30 mm ($1^3/_{16}$ in) across. It also comes from Stoneham, Maine, USA; Písek, Czech Republic; and Iveland, Norway.
Application: the most important Be ore.

Hemimorphite

$Zn_4Si_2O_7(OH)_2 . H_2O$

ORTHORHOMBIC ● ● ●

Properties: C – colorless, white, yellowish, greenish; S – white; L – vitreous to dull; D – transparent to translucent; DE – 3.4; H – 4.5-5; CL – perfect; F – uneven; M – tabular crystals and their combinations, botryoidal and radial aggregates, granular, massive.
Origin and occurrence: Secondary in the oxidation zone of Zn deposits, associated with sphalerite, smithsonite and cerussite. Crystals up to 100 mm (4 in) long come from Bisbee, Arizona, USA. It is also from the El Potosi Mine, Santa Eulalia, Chihuahua, Mexico; Bleiberg, Austria; and Cho-Dien, Vietnam.
Application: Zn ore.

Hemimorphite, 30 mm xx, Mapimi, Mexico

Lamprophyllite

$Na_2Sr_2Ti_3Si_4O_{16}(OH,F)_2$

MONOCLINIC ● ●

Properties: C – brown to dark brown; S – white; L – vitreous to submetallic; D – translucent; DE – 3.5; H – 2-3; CL – perfect; F – uneven; M – tabular crystals, radial aggregates.

Origin and occurrence: Magmatic in alkaline syenites and their pegmatites, associated with nepheline, aegirine and eudialyte. The best crystals up to 150 mm (6 in) long come from Mount Flora, Lovozero massif, Kola Peninsula, Russia. It is also known from Langesundsfjord, Norway and Mont St. Hilaire, Quebec, Canada.

Hemimorphite, 26 mm, Arizona, U.S.A.

Lamprophyllite, 50 mm xx, Lovozero Massif, Kola, Russia

Clinozoisite, 36 mm, Eden Mills, U.S.A.

Clinozoisite

$Ca_2Al_3Si_3O_{12}(OH)$

MONOCLINIC ● ● ● ●

Properties: C – colorless, yellowish, green, pink; S – white; L – vitreous to dull; PS – transparent to translucent; DE – 3.4; H – 6.5; CL – good; F – uneven; M – prismatic crystals, columnar, radial aggregates, granular, massive.

Origin and occurrence: Metamorphic and hydrothermal in contacts of marbles and in the Alpine-type veins. Prismatic crystals up to 60 mm (2³/₈ in) long come from Radoy Island, Norway. It comes also from Eden Mills, Vermont; Allens Park, Colorado, USA; Pinos Altos, Baja California, Mexico.

Epidote

$Ca_2(Al,Fe^{3+})_3Si_3O_{12}(OH)$

MONOCLINIC ● ● ● ● ●

Properties: C – green, brown, greenish, yellow-green; S – white; L – vitreous to dull; D – transparent to translucent, locally opaque; DE – 3.4; H – 6-7; CL – good; F – uneven; M – prismatic crystals, columnar and radial aggregates, granular, massive.

Origin and occurrence: Metamorphic in marble/granite contacts; hydrothermal in the Alpine-type veins and in hydrothermally altered rocks, typically associated with albite, prehnite and amphibole. Its green crystals up to 140 mm (5½ in) long, were found in Sobotín, Czech Republic. Perfect crystals up to 100 mm (4 in) across come from Knappenwand, Austria. Thick tabular crystals occur in Prince of Wales Island, Alaska, USA. Fine crystals, resembling Austrian crystals, were found recently in Alchuri, Shigar, Pakistan. Fine columnar aggregates of crystals are known from Pampa Blanca, Peru and also reported from Arendal, Norway.

Piemontite

$Ca_2(Al,Mn^{3+})_3Si_3O_{12}(OH)$

MONOCLINIC ● ●

Properties: C – red-brown to black, crimson, red-yellow; S – white; L – vitreous to dull; D – translucent, locally opaque; DE – 3.5; H – 6; CL – good; F – uneven; M – prismatic crystals, radial aggregates, granular.

Origin and occurrence: Metamorphic in shales and Mn-rich metamorphic rocks; rare magmatic in rhyolites and pegmatites. Needles up to 30 mm (1³/₁₆

Epidote, Dashkesan, Azerbaidzhan

Epidote, 50 mm, Knappenwand, Austria

in) long come from St. Marcel, Piedmont, Italy and Otakiyama, Japan.

Allanit-(Ce)

$(Ca,Ce,Y)_2(Al,Fe^{3+})_3Si_3O_{12}(OH)$

MONOCLINIC ● ● ●

Properties: C – black to dark brown; S – light gray; L – greasy to submetallic; D – translucent to opaque; DE – 3.9; H – 5.5-6; CL – none; F – conchoidal to uneven; M – tabular crystals, granular; R – usually metamict.

Origin and occurrence: Magmatic in pegmatites and granites; metamorphic in various types of metamorphic rocks, e.g. migmatites, amphibolites and gneisses. Grains, up to 70 cm (27$^{9}/_{16}$ in) across found in pegmatites near Bancroft, Ontario, Canada. Also occurs in Barringer Hill, Colorado; Amelia district, Virginia, USA; Arendal and Hitterö, Norway; Ytterby and Riddarhyttan, Sweden; and Yates mine, Quebec, Canada.

Piemontite, 45 mm, Todyryact, Morocco

Allanite-(Ce), 40 mm, Vizcaya, Spain

Tanzanite, 37 mm, Arusha, Tanzania

Zoisite, 60 mm, Weisenstein, Germany

Zoisite

$Ca_2Al_3Si_3O_{12}(OH)$

ORTHORHOMBIC ● ● ●

Varieties: thulite, tanzanite

Properties: C – colorless, yellowish, green, pink, red (thulite), blue (tanzanite); S – white; L – vitreous to dull; D – transparent to translucent; DE – 3.4; H – 6.5; CL – good; F – uneven; M – prismatic crystals, radial aggregates, granular.
Origin and occurrence: Metamorphic in regionally metamorphosed Ca-rich rocks, mainly in pyroxenic gneisses, amphibolites and in marble contacts. It is known from many localities like Saualpe, Austria; Zermatt, Switzerland; Lexviken, Norway (thulite); Traversella, Italy; Alchuri, Pakistan; Merelani Hills, Arusha, Tanzania (tanzanite), where crystals up to 70 mm (2¾ in) long were found.
Application: tanzanite and thulite are cut as gemstones.

Thulite, 40 mm, Lexviken, Norway

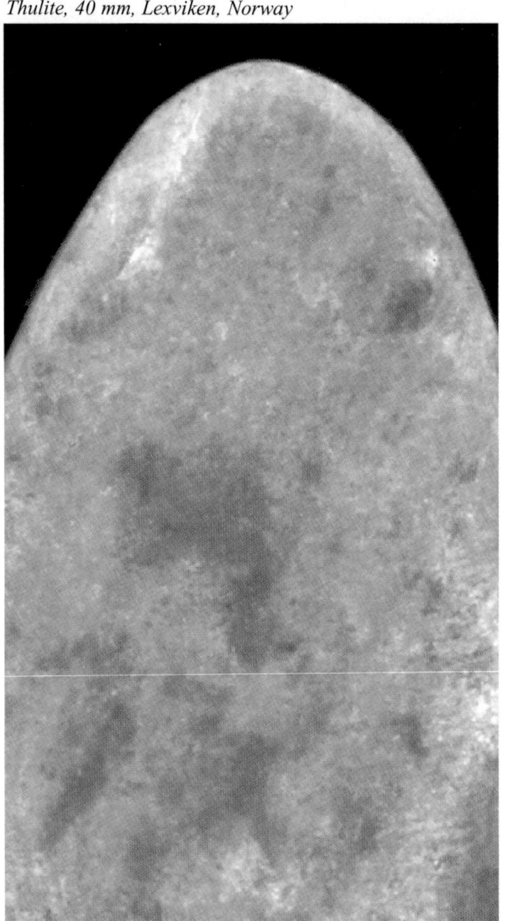

Vesuvianite

$Ca_{19}(Al,Mg,Fe)_{13}Si_{18}O_{68}(OH,F,O)_{10}$

TETRAGONAL ● ● ●

Properties: C – brown, yellowish, green, blue, purple, colorless; S – white; L – vitreous to dull; D – transparent to translucent; DE – 3.3; H – 6-7; CL – imperfect; F – uneven to conchoidal; M – prismatic to tabular crystals, columnar aggregates with radial structure, granular, massive.
Origin and occurrence: Metamorphic and hydrothermal in contact metamorphosed Ca-rich rocks, mainly in skarns, in marble contacts, also in rodingites; rarely magmatic in alkaline rocks. It is usually associated with grossular, wollastonite and diopside. Perfect green crystals up to 180 mm (7 1/16 in) long and purple crystals up to 70 mm (2¾ in) long come from the Jeffrey quarry, Asbestos, Quebec, Canada. It is also known from Hazlov, Czech Republic; Crestmore, California, Franklin, New Jersey, USA; and Monzoni, Italy.

Vesuvianite, 110 mm, Asbestos, Canada

Viluite

$Ca_{19}(Al,Mg)_{13}B_5Si_{18}O_{68}(O,OH)_{10}$

TETRAGONAL ●

Properties: C – dark green, gray-brown; S – white; L – vitreous to dull; D – transparent to translucent; DE – 3.4; H – 6; CL – imperfect; F – uneven to conchoidal; M – prismatic crystals.

Origin and occurrence: Metamorphic in serpentinized skarn, associated with grossular. Its perfect prismatic crystals up to 50 mm (2 in) long are only known from the Vilui River basin, Yakutia, Russia.

Viluite, 30 mm xx, Vilui River, Russia

Benitoite

$BaTiSi_3O_9$

TRIGONAL ● ●

Properties: C – blue, pink, white, colorless; S – white; L – vitreous to dull; D – transparent to translucent; DE – 3.6; H – 6-6.5; CL – imperfect; F – uneven to conchoidal; M – prismatic to tabular crystals, mainly with trigonal cross-section; LU – bluish.
Origin and occurrence: Hydrothermal in veins, cross-cutting serpentinites, always associated with neptunite and natrolite. Its classic locality is the Benitoite Gem mine, San Benito Co., California, USA, where it forms tabular crystals up to 40 mm (1 9/16 in) across.

Catapleiite

$Na_2ZrSi_3O_9 . 2 H_2O$

HEXAGONAL ● ●

Properties: C – light yellow, yellow-brown, pink, brown, blue; S – white; L – vitreous to dull; D – transparent to opaque; DE – 2.8; H – 5-6; CL – good; F – uneven; M – thin tabular crystals, lamellar aggregates.
Origin and occurrence: Magmatic in nepheline syenites and their pegmatites, together with aegirine, titanite, nepheline and microcline. It occurs in Mont St.-Hilaire, Quebec, Canada, as tabular crystals up to 150 mm (6 in) across. Crystals up to 30 mm (1 3/16 in) across come from Mount Yukspor, Khibiny massif, Kola Peninsula, Russia. It is also known from Langesundsfjord, Norway and Magnet Cove, Arkansas, USA.

Eudialyte

$Na_4(Ca,Fe,Ce,Mn)_2ZrSi_6O_{18}(OH,Cl)$

TRIGONAL ● ●

Properties: C – red, pink to brown; S – white; L – vitreous to dull; D – transparent to translucent; DE – 2.8; H – 5-5.5; CL – imperfect; F – uneven; M – prismatic and tabular crystals, granular.
Origin and occurrence: Magmatic in nepheline syenites and their pegmatites, associated with aegirine, nepheline and microcline. Crystals up to 80 mm (3 1/8 in) across come from Mount Kukisvumchorr, Khibiny massif, Kola Peninsula, Russia. Crystals up to 50 mm (2 in) across are also known from Mont St.-Hilaire, Quebec, Canada. It also occurs in Langesundsfjord, Norway and in Los Island, Guinea.

Ferroaxinite

$Ca_2FeAl_2BSi_4O_{15}(OH)$

TRICLINIC ● ● ●

Properties: C – brown to purple- brown, light purple; S – white; L – vitreous to dull; D – trans-

Benitoite, 28 mm x, San Benito Co., U.S.A.

Catapleiite, 37 mm, Mont St.-Hilaire, Canada

Eudialyte, 20 mm grain, Khibiny Massif, Kola, Russia

Ferroaxinite, 62 mm, Khapalu, Pakistan

parent to translucent; DE – 3.3; H – 6. 5-7; CL – good; F – uneven to conchoidal; M – tabular crystals, platy aggregates, granular, massive.
Origin and occurrence: Metamorphic and hydrothermal in contacts of marbles and granites, associated with clinozoisite, prehnite, calcite and actinolite, also in the Alpine-type veins and pegmatites. Perfect tabular crystals up to 150 mm (6 in) across come from Puiva, Polar Ural, Russia. Other renowned localities are Obira, Japan; Bourg d'Oisans, France; and Monte Scopi, Switzerland.

Tinzenite
$Ca(Mn,Fe)_2Al_2BSi_4O_{15}(OH)$

TRICLINIC ● ●

Properties: C – yellow, orange to red; S – white; L – vitrous to dull; D – transparent to translucent; DE – 3.3; H – 6.5-7; CL – good; F – uneven to conchoidal; M – tabular crystals, platy and fibrous aggregates, granular, massive.
Origin and occurrence: Hydrothermal in the Alpine-type veins, cross-cutting a rock, rich in braunite. It comes from Tinzen, Val d'Err, Switzerland and in the Cassagna mine, Genova, Italy.

Tinzenite, 10 mm xx, Val Graveglia, Italy

Beryl

$Be_3Al_2Si_6O_{18}$

HEXAGONAL ● ● ● ●

Varieties: emerald, aquamarine, heliodor, morganite, goshenite, red beryl (bixbite).

Properties: C – variable in different varieties; common beryl – mostly yellow, yellow-green, light green to white, rare blue; varieties: emerald – dark emerald-green; aquamarine – light to dark blue-green; morganite – pink; heliodor – light yellow to yellow-green; goshenite – white to colorless; red beryl B red; S – white; L – vitreous to dull; D – transparent to translucent; DE – 2.6; H – 7.5-8; CL – imperfect; F – uneven to conchoidal; M – long prismatic to tabular crystals, columnar and radial aggregates, granular, massive.

Origin and occurrence: Magmatic in pegmatites and granites; hydrothermal in greisens, in cavities in rhyolite, in quartz veins; metamorphic in mica schists. Perfect prismatic crystals of common beryl up to 9 m (29 ft 6 in) long found in the Etta mine, Keystone, South Dakota, USA. Crystals weighing up to 177 tons, come from Namivo, Alto Ligonha, Mozambique. Other localities are Pici, Brazil; Iveland, Norway; and Antsirabé, Madagascar. Emerald crystals occur in mica-schists, marbles and ultrabasic rocks, associated with other Be minerals, phenakite

Emerald, 30 mm x, Boyacá, Colombia

Emerald, 30 mm, Ural Mts., Russia

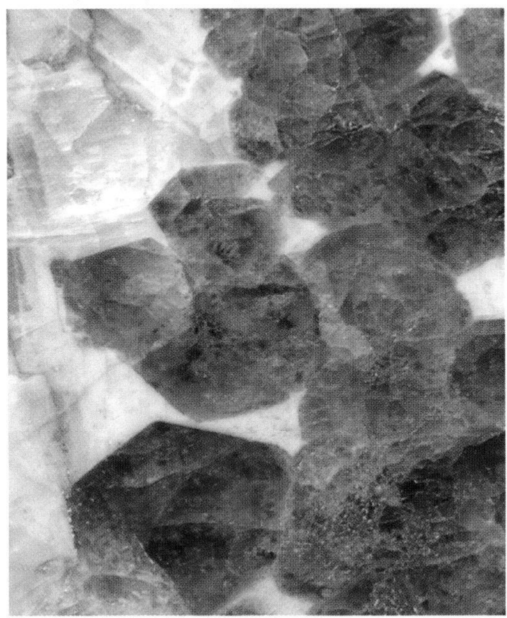

and chrysoberyl. Beautiful dark green transparent crystals are known from Malyshevo, Ural Mts., Russia; Muzo and Coscuez, Colombia; Habachtal, and

Emerald, 90 mm, Muzo, Colombia

Aquamarine, 97 mm, Koronel Hurta, Brazil

Austria, where the largest crystals reach up to 120 mm ($4^{11}/_{16}$ in) in size. Aquamarine is mainly known from pegmatites and hydrothermal veins, commonly

Aquamarine, 158 mm, Shingus, Pakistan

Heliodor, 87 mm, Volodarsk Volynskii, Ukraine

associated with tourmaline, quartz and albite. Intense blue crystals are known from many pegmatites in Minas Gerais, Brazil, where in the Marambaia mine, crystals, up to 70 cm ($27\frac{9}{16}$ in) long and weighing up to 110 kg (242 lb) found. Very beautiful gemmy crystals over 30 cm (12 in) long recently occurred in the Medina Mine. It also come from Murzinka, Ural mountains and Adun Chilon, Siberia, Russia. It also comes from Spitzkopje, Namibia; Gilgit, Pakistan and elsewhere. Morganite is a typical mineral of granitic pegmatites, where it occurs in mainly in cavities, usually associated with color varieties of tourmaline, quartz and albite. Its tabular crystals up to 100 mm (4 in) across found in the White Queen mine, Pala, California, USA. Smaller crystals come from San Piero in Campo, Elba, Italy. Crystals up to 50 cm (20 in) across reported from several localities in Minas Gerais, Brazil. Heliodor also occurs in pegmatite cavities, hydrothermal veins and in gneisses, commonly associated with quartz and albite. Prismatic crystals up to 200 mm ($7\frac{7}{8}$ in) long, come from Volodarsk Volynskii, Ukraine. It is also known from Nerchinsk, Siberia, Russia and several mines in Minas Gerais, Brazil. Goshenite occurs only in pegmatites. Its prismatic crystals come from Goshen, Massachusetts,

Heliodor, 26 mm x, Pamir, Tadzhikistan

Goshenite, 39 mm, Apalygun, Pakistan

USA, and San Piero in Campo, Elba, Italy. Red beryl occurs in Violet Claims, Wah Wah mountains, Utah, USA, where crystals up to 50 mm (2 in) were found. *Application:* Be ore, color varieties are cut as gemstones.

Red beryl, 17 mm x, Wah Wah Mts., U.S.A.

Morganite, 43 mm x, San Diego Co., U.S.A.

Bazzite, 70 mm, Tordal, Norway

Sekaninaite, 40 mm x, Dolní Bory, Czech Republic

Bazzite
$Be_3(Sc,Al)_2Si_6O_{18}$

HEXAGONAL ●

Properties: C – light to intense blue; S – white; L – vitreous to dull; D – transparent to translucent; DE – 2.8; H – 6.5; CL – imperfect; F – uneven to conchoidal; M – long prismatic crystals, columnar and radial aggregates.
Origin and occurrence: Hydrothermal in the Alpine-type veins and pegmatites. Crystals up to 20 mm (25/32 in) long come from Tørdal, Norway. It is also known from Lago Maggiore, Italy and St. Gotthard, Switzerland.

Properties: C – blue, purple, gray, gray-green, gray-brown, trichroic; S – white; L – vitreous to dull; D – transparent to translucent; DE – 2.5; H – 7-7.5; CL – imperfect; F – uneven to conchoidal; M – short prismatic crystals, granular.
Origin and occurrence: Metamorphic in migmatites and gneisses and in contact cherts; magmatic in granites and granitic pegmatites, usually associated with andalusite and sillimanite, also known from placers. It is a typical rock-forming mineral. Its prismatic crystals are very rare. Poorly developed transparent crystals up to 200 mm (7 7/8 in) long come from Näverberg, Sweden. It also occurs in Orijärvi, Finland; Kragerö, Norway; and Bodenmais, Germany. Gemmy pebbles are known from the vicinity of Ratnapura, Sri Lanka.

Cordierite
$Mg_2Al_4Si_5O_{18}$

ORTHORHOMBIC ● ● ● ●

Varieties: iolite (gemmy blue)

Sekaninaite
$Fe_2Al_4Si_5O_{18}$

ORTHORHOMBIC ● ● ●

Properties: C – blue, purple, strongly pleochroic; S

Cordierite, 50 mm, Fishtail Lake, Canada

Dravite, 27 mm, Gujarkkot, Nepal

– white; L – vitreous to dull; D – transparent to translucent; DE – 2.8; H – 7-7.5; CL – imperfect; F – uneven to conchoidal; M – short prismatic crystals, granular.
Origin and occurrence: Magmatic in granitic pegmatites and some granites, associated with andalusite and tourmaline; metamorphic in gneisses and migmatites. Conical, imperfect crystals, up to 70 cm ($27^{9}/_{16}$ in) long typically come from Dolní Bory, Czech Republic; also known from San Piero in Campo, Elba, Italy.

Dravite

TOURMALINE GROUP
$NaMg_3Al_6(BO_3)_3Si_6O_{18}(OH)_4$

TRIGONAL ● ● ● ●

Properties: C – light to dark black-brown, blue, colorless, commonly pleochroic; S – white; L – vitreous to dull; D – transparent to translucent; DE – 3.0; H – 7-7.5; CL – none; F – uneven to conchoidal; M – long to short prismatic crystals, columnar to acicular aggregates, granular.
Origin and occurrence: Metamorphic in migmatites, gneisses, mica schists, marbles and in contact metasomatic rocks; magmatic in some granitic pegmatites; hydrothermal in quartz veins and ore veins, also known from placers. Brown to dark brown perfect crystals up to 200 mm ($7^{7}/_{8}$ in) long come from mica schists near Dravograd, Slovenia. It is also known from marbles in Gouverneur, New York, USA. Crystals up to 150 mm (6 in) across found in Yinnietharra, Western Australia. Crystals up to 50 mm (2 in) long were recently found in Gujarkot, Nepal.
Application: transparent crystals are cut as gemstones.

Buergerite, 12 mm x, San Luis Potosí, Mexico

Buergerite

TOURMALINE GROUP
$NaFe^{3+}{}_3Al_6(BO_3)_3Si_6O_{21}F$

TRIGONAL ●

Properties: C – black, strongly pleochroic; S – yellow-brown; L – vitreous to dull; D – translucent to opaque; DE – 3.3; H – 7; CL – none; F – uneven to conchoidal; M B prismatic crystals, granular.
Origin and occurrence: Hydrothermal in rhyolites. Its black crystals, up to 40 mm (19/16 in) long, are known from Mexquitic, San Luis Potosi, Mexico.

Schorl, 40 mm, Gilgit, Pakistan

Schorl

TOURMALINE GROUP

$NaFe_3Al_6(BO_3)_3Si_6O_{18}(OH)_4$

TRIGONAL ● ● ● ●

Properties: C – black, black-brown, blue-black, strongly pleochroic; S – white; L – vitreous to dull; D – translucent to opaque; DE – 3.3; H – 7-7.5; CL – none; F – uneven to conchoidal; M – long to short prismatic crystals, columnar to acicular aggregates, granular, massive.

Origin and occurrence: Magmatic in granites and granitic pegmatites; hydrothermal in greisens, in quartz and ore veins; metamorphic in migmatites, gneisses, mica schists and tourmalinites; also known from placers. It is usually associated with muscovite, quartz and albite. Perfect black crystals come from many pegmatite localities. Its long prismatic crystals, up to 5 m long, come from Arendal, Norway. Very good crystals are also known from Kaatiala, Finland; Dolni Bory, Czech Republic; Conselheira Pena and Galileia, Minas Gerais, Brazil.

Povondraite

TOURMALINE GROUP

$NaFe^{3+}_3Mg_2Fe^{3+}_4(BO_3)_3Si_6O_{18}(OH)_4$

TRIGONAL ●

Properties: C – black; S – gray; L – vitreous to dull; D – translucent to opaque; DE – 3.3; H – 7; CL –

Povondraite, 60 mm, Alto Chapare, Bolivia

Elbaite, 115 mm, Tourmaline Queen Mine, U.S.A.

Elbaite, 153 mm, Afghanistan

none; F – uneven to conchoidal; M – short prismatic crystals, granular.
Origin and occurrence: Hydrothermal along the cracks in metamorphosed evaporites. Its black crystals up to 10 mm (3/8 in) long typically come from Alto Chapare, Cochabamba, Bolivia.

Elbaite

TOURMALINE GROUP

$Na(Li_{1,5}Al_{1,5})Al_6(BO_3)_3Si_6O_{18}(OH)_3F$

TRIGONAL ● ● ●

Varieties: rubellite, verdelite, indiccolite, achroite

Properties: C – varies in different varieties, rubellite – pink to red; verdelite – various hues of green; indiccolite – blue; achroite – colorless, other colors include yellow, brown and black; S – white; L – vitreous to dull; D – transparent to translucent; DE – 3.0; H – 7; CL – none; F – uneven to conchoidal; M – prismatic crystals, columnar aggregates, granular, massive.
Origin and occurrence: Almost only magmatic and hydrothermal in granitic pegmatites, typically associated with lepidolite and albite; also in placers. Its crystals are mostly known from pegmatite

Rubellite, 23 mm, Pala, U.S.A.

cavities in many localities. Rubellite crystal called 'The Rocket' (109 cm/43$^5/_{16}$ in long) was found in the Jonas mine, Minas Gerais, Brazil. Rubellite crystals up to 40 cm (15¾ in) long come from Alto Ligonha district, Mozambique. Rubellite crystals up to 250 mm (9$^{13}/_{16}$ in) long occurred in the Stewart Lithia, Tourmaline King and Tourmaline Queen mines, Pala, California, USA. Verdelite crystal 270 mm (10$^5/_8$ in) long, is known from the Dunton mine, Newry, Maine, USA. Other important localities include San Piero in Campo, Elba, Italy; Malkhan, Transbaikalia, Russia; Rožná , Czech Republic; Utö, Sweden; Gilgit, Pakistan; and Paprok, Afghanistan.
Application: commonly used as a gemstone.

Uvite

TOURMALINE GROUP
$CaMg_4Al_5(BO_3)_3Si_6O_{18}(OH,F)_4$

TRIGONAL ● ● ●

Properties: C – gray, black, brown, green, red, pleochroic; S – white; L – vitreous to dull; D – translucent to opaque; DE – 3.3; H – 7-7.5; CL – none; F – uneven to conchoidal; M – long to short prismatic crystals, columnar aggregates, granular.
Origin and occurrence: Metamorphic in Ca-rich rocks, marbles, skarns; magmatic in some pegmatites; hydrothermal in ore veins. Perfect green and red crystals, up to 30 mm (1$^3/_{16}$ in) across, occur in Brumado, Bahía, Brazil. It is also known from Gouverneur and Pierrepont, New York, USA.

Uvite, 41 mm, Brumado, Brazil

Liddicoatite

TOURMALINE GROUP
$Ca(Li_2Al)Al_6(BO_3)_3Si_6O_{18}(OH)_3F$

TRIGONAL ● ● ●

Varieties: rubellite, verdelite

Properties: C – varies in different varieties, mainly pink, green, green-brown to yellow-brown; S – white; L – vitreous to dull; D – transparent to translucent; DE – 3.1; H – 7; CL – none; F – uneven to conchoidal; M – prismatic crystals, columnar aggregates, granular.
Origin and occurrence: Magmatic in granitic peg-

Liddicoatite, 60 mm, Anjanobonoina, Madagascar

Verdelite, 24 mm, Gillette Quarry, U.S.A.

matites in association with lepidolite, spodumene and albite. Its perfect red and green crystals up to 250 mm ($9^{13}/_{16}$ in) long come from pegmatite cavities in many localities in Madagascar, e.g. Sahatany and

Indigolite, 32 mm, Esmeralda Mine, Mesa Grande, U.S.A.

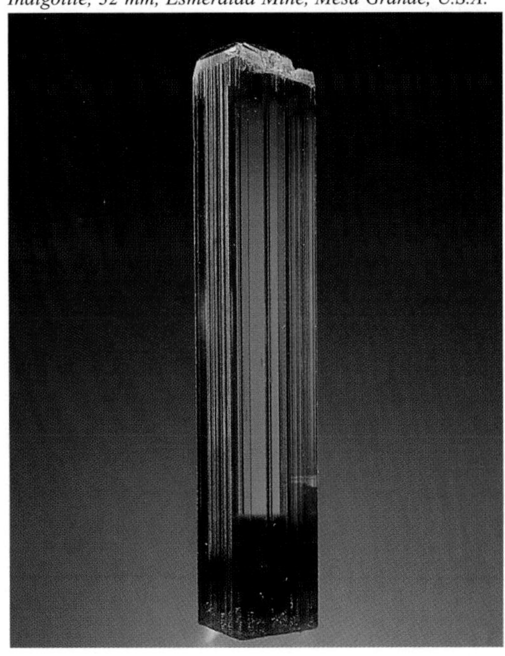

Anjanabonoina. It is also known from Bližná, Czech Republic.
Application: cut as a gemstone.

Liddicoatite, 41 mm, Madagascar

Foitite

TOURMALINE GROUP

□ $Fe_2AlAl_6(BO_3)_3Si_6O_{18}(OH)_4$

TRIGONAL ● ● ●

Properties: C – black to black-purple; S – white; L – vitreous to dull; D – translucent to opaque; DE – 3.3; H – 7; CL – none; F – uneven to conchoidal; M – prismatic crystals, acicular aggregates.
Origin and occurrence: Magmatic in granitic pegmatites and granites; hydrothermal in pegmatite cavities. Foitite typically forms black tips of elbaite crystals (moors heads) in Dobrá Voda, Czech Republic; San Piero in Campo, Elba, Italy; White Queen mine, Pala, California, USA. It occurs together with schorl in some granitic pegmatites, like Rožná, Czech Republic.

Rossmanite

TOURMALINE GROUP

□ $LiAl_2Al_6(BO_3)_3Si_6O_{18}(OH)_4$

TRIGONAL ●

Properties: C – pink; S – white; L – vitreous to dull; D – transparent to translucent; DE – 3.1; H – 7; CL – none; F – uneven to conchoidal; M – prismatic crystals, columnar aggregates.
Origin and occurrence: Magmatic in granitic pegmatites. Prismatic crystals, up to 20 mm ($^{25}/_{32}$ in) long, were found in massive lepidolite in Rožná, also in Lastovicky, Czech Republic.

Dioptase

$Cu_6Si_6O_{18} \cdot 6\,H_2O$

TRIGONAL ● ●

Properties: C – emerald-green to blue-green; S – light blue-green; L – vitreous to dull; D – transparent to translucent; DE – 3.3; H – 5; CL – good; F – uneven to conchoidal; M – long to short prismatic crystals, granular.
Origin and occurrence: Secondary in the oxidation zone of Cu deposits, associated with other secondary Cu minerals. Beautiful crystals up to 50 mm (2 in) across occur in Tsumeb, Namibia; Altyn Tyube, Kazakhstan; Renneville, Congo. It also comes from the Mammoth mine, Tiger, Arizona, USA.

Milarite

$KCa_2AlBe_2Si_{12}O_{30} \cdot H_2O$

HEXAGONAL ● ●

Properties: C – colorless, white, gray, light green; S – white; L – vitreous to dull; D – transparent to translucent; DE – 2.5; H – 5-6; CL – none; F – uneven to conchoidal; M – long prismatic to acicular crystals, granular.
Origin and occurrence: Hydrothermal in the Alpine-type veins, in pegmatites and hydrothermal veins, associated with adularia. Crystals up to 40 mm ($1^{9}/_{16}$ in) long come from Jaguaraçu, Minas Gerais, Brazil. It was also found in St. Gotthard, Switzerland; Klein Spitzkopje, Namibia; and Valencia mine, Guanajuato, Mexico.

Foitite, 20 mm x, Dobrá Voda, Czech Republic

Rossmanite, 10 mm xx, Laštovičky, Czech Republic

Dioptase, 40 mm, Mindouli, Zair

Sugilite

$KNa_2(Fe,Mn,Al)_2Li_3Si_{12}O_{30}$

HEXAGONAL ●

Properties: C – purple; S – white; L – vitreous to dull; D – transparent to translucent; DE – 2.7; H – 6-6.5; CL – none; F – uneven to conchoidal; M – long prismatic to acicular crystals, granular.
Origin and occurrence: Hydrothermal in alkaline syenites, associated with pectolite, albite and aegirine. It occurs in Iwagi Island, Japan and Hotazel, South Africa. *Application:* cut and polished as a gemstone and decorative stone.

Milarite, 22 mm, Jaguaracu, Brazil

Sugilite, 40 mm, Hotazel, South Africa

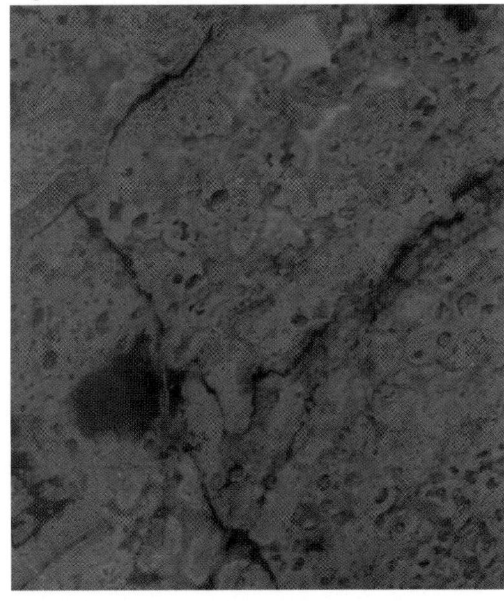

Bronzite, 15 mm, Minas Gerais, Brazil

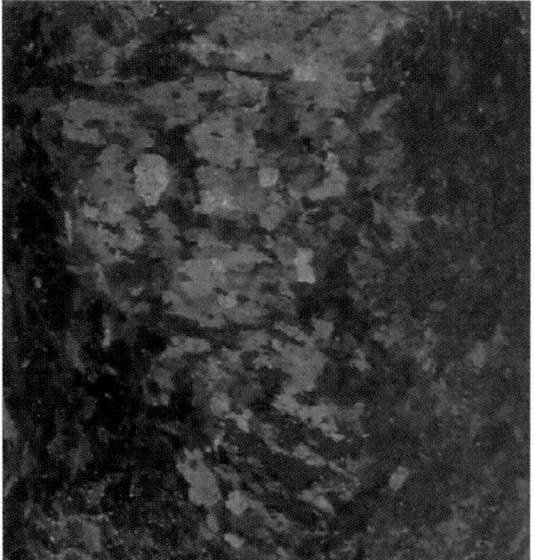

Chrome diopside, 5 mm grains, Inagli, Russia

Enstatite

PYROXENE GROUP

$Mg_2Si_2O_6$

ORTHORHOMBIC ● ● ● ●

Varieties: bronzite

Properties: C – colorless, gray, yellowish, greenish, brown (bronzite); S – white; L – vitreous to dull; D – transparent to opaque; DE – 3.2; H – 5-6; CL – good; F – uneven to conchoidal; M – prismatic crystals, granular.

Origin and occurrence: Magmatic in ultrabasic rocks, gabbros, peridotites and in meteorites; metamorphic in marbles, associated mainly with olivine and pyrope. Known from Bamle, Norway in crystals up to 50 cm (20 in) long. It was originally described from Ruda nad Moravou, Czech Republic. Gemmy crystals up to 20 mm (25/32 in) long occur in Brumado, Bahía, Brazil.

Diopside

PYROXENE GROUP

$CaMgSi_2O_6$

MONOCLINIC ● ● ● ● ●

Varieties: chrome diopside, fassaite, jeffersonite

Properties: C – light green, dark green (chrome diopside), colorless, white, gray, brown, rare blue; S – white; L – vitreous to dull; D – transparent to translucent; DE – 3.3; H – 5.5-6.5; CL – good; F – uneven to conchoidal; M – long to short prismatic crystals, granular.

Origin and occurrence: Metamorphic in Ca-rich rocks, skarns, pyroxene gneisses, marbles; magmatic in basic igneous rocks, pegmatites and meteorites; hydrothermal in the Alpine-type veins. Fassaite occurs in skarns, chrome diopside in metamorphic deposits of Cu and Cr; jeffersonite in metamorphic deposits of Mn and Zn. Diopside is a typical rock-forming mineral,

Diopside, 22 mm x, Ala, Italy

associated with plagioclase, grossular and epidote. Well formed crystals are relatively rare. Jeffersonite forms prismatic crystals up to 250 mm (9 13/16 in) long in Franklin, New Jersey, USA. Crystals in Corrego Setuba, Minas Gerais, Brazil reach up to 30 cm (12 in). Other typical localities are Zillertal, Austria; Nordmarken, Sweden; Orford mine, Quebec, Canada. Fassaite is known from Val di Fassa, Italy. Chrome diopside crystals reach up to 100 mm (4 in) in Outokumpu, Finland and gem rough recently found in Inagli, Yakutia, Russia.

Hedenbergite

PYROXENE GROUP

$CaFeSi_2O_6$

MONOCLINIC ● ● ● ●

Properties: C – dark green, brown-green, brown to black; S – white to gray; L – vitreous to dull; D – transparent to translucent; DE – 3.6; H – 6; CL – good; F – uneven to conchoidal; M – long to short prismatic crystals, granular.

Origin and occurrence: Metamorphic in Ca-rich rocks, like Fe-skarns and pyroxene gneisses; magmatic in some granites and syenites, associated with magnetite, grossular and epidote. Crystals are relatively rare, reaching up to 50 mm (2 in) from Dalnegorsk, Russia and Franklin, New Jersey, USA. Also occurs in Nordmarken, Sweden. Large lamellar aggregates come from skarns in Rio Marina, Elba, Italy.

Hedenbergite, 40 mm, Dalnegorsk, Russia

Augite, 17 mm, Firmerich, Germany

Augite

PYROXENE GROUP

$(Ca,Mg,Fe,Al)(Si,Al)_2O_6$

MONOCLINIC ● ● ● ● ●

Properties: C – dark brown to black; S – gray-green; L – vitreous to dull; D – translucent to opaque; DE – 3.6; H – 6; CL – good; F – uneven to conchoidal; M – short prismatic crystals, granular.

Origin and occurrence: Magmatic in basic rocks (basalts, gabbros, diabases and their tuffs); rare metamorphic in skarns. It is a typical rock-forming mineral, known from many localities. Well formed crystals up to 150 mm (6 in) across occur mainly in volcanic rocks, such as near Lake Clear, Ontario, Canada. Also known from Laacher See, Germany; Lukov and Paskapole, Czech Republic, in crystals, up to 50 mm (2 in) in size.

Omphacite, 90 mm, St.Leonhard, Austria

Omphacite

PYROXENE GROUP

$(Ca,Na)(Mg,Fe^{2+},Fe^{3+},Al)(Si,Al)_2O_6$

MONOCLINIC ● ● ● ●

Properties: C – green to dark green; S – gray-white; L – vitreous to dull; D – translucent to opaque; DE – 3.3; H – 5-6; CL – good; F – uneven to conchoidal; M – short prismatic crystals, acicular aggregates, granular. *Origin and occurrence:* Metamorphic in ultrabasic and basic rocks, originated under high-temperature, usually associated with pyrope, diopside and kyanite. It occurs in eclogites and granulites in Rubinberg, Germany; Headsburg, California, USA and elsewhere.

Jadeite, 85 mm, Guerro Negro, Baja California, Mexico

Nephrite, 60 mm, Jordanów, Poland

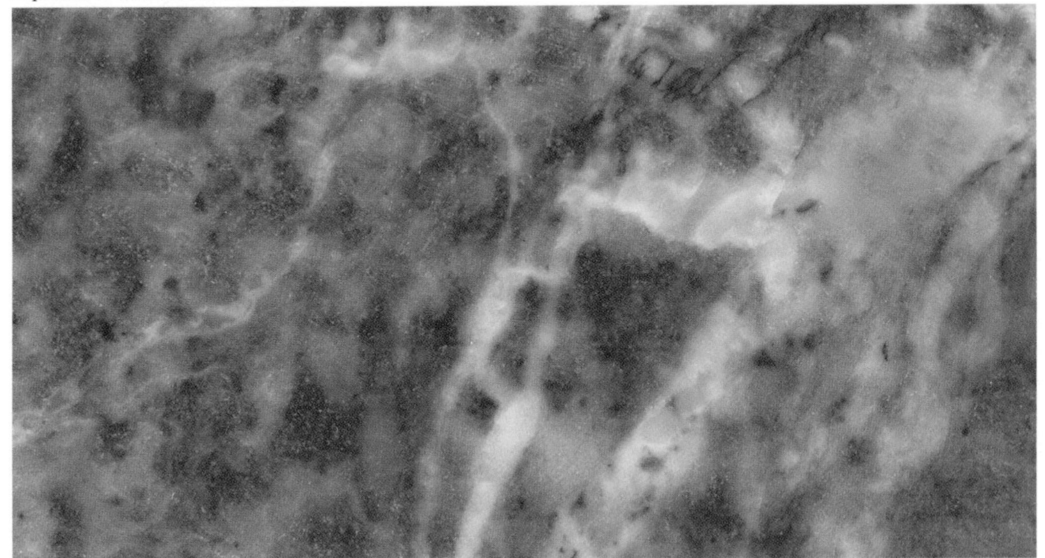

Jadeite

PYROXENE GROUP

$NaAlSi_2O_6$

MONOCLINIC ● ●

Varieties: nephrite

Properties: C – white, lavender to gray, in aggregates also light green (nephrite); S – white; L – vitreous to dull; D – transparent to translucent; DE – 3.2; H – 6; CL – good; F – uneven to conchoidal; M – prismatic crystals, acicular aggregates, granular.
Origin and occurrence: Exclusively metamorphic in strongly metamorphosed rocks; also in placers. Blocks, weighing several tons, are known from Ben Sur, California, USA. It also come from Tawmaw, Burma; New Zealand; Tibet and elsewhere.
Application: as a material for carvings and decorative purposes.

Aegirine

PYROXENE GROUP

$NaFe^{3+}Si_2O_6$

MONOCLINIC ● ● ●

Properties: C – dark green to black-green; S – light yellow-gray; L – vitreous to dull; D – translucent to opaque; DE – 3.6; H – 6; CL – good; F – uneven to conchoidal; M – long prismatic to acicular crystals, acicular aggregates.
Origin and occurrence: Magmatic, mainly in alkaline magmatic rocks (syenites, carbonatites, alkaline granites and their pegmatites); rare hydrothermal in sediments. Well formed prismatic crystals up to 150 mm (6 in) long come from Mount Malosa, Malawi. Crystals up to 30 cm (12 in) occur in Langensundsfjord, Norway. Other localities are Mount Karnasurt, Lovozero massif, Kola Peninsula, Russia and Mont St.-Hilaire, Quebec, Canada.

Aegirine, 77 mm, Mt.Malosa, Malawi

Spodumene, 52 mm x, Kunar, Afghanistan

Hiddenite, 35 mm, Adams Property, U.S.A.

Kunzite, 118 mm, Nuristan, Afghanistan

Spodumene

PYROXENE GROUP

$LiAlSi_2O_6$

MONOCLINIC ● ● ●

Varieties: kunzite, hiddenite, triphane

Properties: C – white, gray, yellowish (triphane), green (hiddenite), pink to purple (kunzite); S – white; L – vitreous to dull; D – transparent to translucent; DE – 3.2; H – 6.5-7.5; CL – good; F – uneven to conchoidal; M – long to short prismatic crystals, cleavable aggregates, granular; LU – orange.
Origin and occurrence: Magmatic in granitic pegmatites; hydrothermal in pegmatite cavities. It is known from many localities. Its poorly developed crystals up to 6 m (20 ft) long occur in the Etta mine, Keystone, South Dakota and in Kings Mountain, North Carolina, USA. Transparent kunzite crystals up to 40 cm (15¾ in) long come from Mawi, Laghman, Afghanistan. Kunzite crystals up to 280 mm (11 in) long found in the Pala Chief mine, Pala, California, USA. Cracked kunzite crystals, up to 1 m (39⅜ in) long known from Araçuai, Minas Gerais, Brazil. Hiddenite occurs in crystals up to 250 mm (9 13/16 in) long in Resplendor, Minas Gerais, Brazil. It also comes from the Adams property, North Carolina, USA.
Application: raw material for ceramics, kunzite and hiddenite are cut as gemstones.

Carpholite, 70 mm, Horní Slavkov, Czech Republic

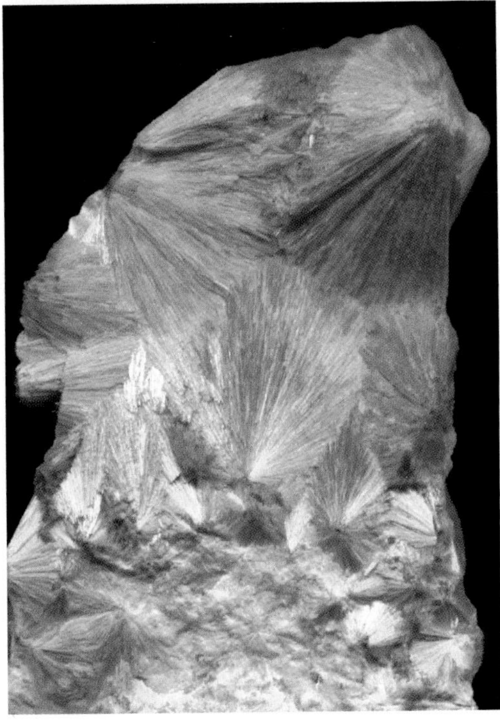

Carpholite

$MnAl_2Si_2O_6(OH)_4$

ORTHORHOMBIC ● ●

Properties: C – various hues of yellow; S – white; L – vitreous to dull; D – translucent; DE – 3.O; H – 5-5.5; CL – good; F – uneven to conchoidal; M – acicular crystals and their aggregates.
Origin and occurrence: Hydrothermal in veins, cross-cutting greisens, associated with fluorite, quartz and cassiterite; also in metamorphic rocks. It was described from Horní Slavkov, Czech Republic, where it forms acicular crystals up to 10 mm (3/8 in) long and radial aggregates. It is also known from Meuville, Belgium and Wippra, Germany.

Lorenzenite, 20 mm xx, Flora Mt., Kola, Russia

Lorenzenite

$Na_2Ti_2Si_2O_9$

ORTHORHOMBIC ● ●

Properties: C – brown to black; S – yellowish; L – vitreous to dull; D – translucent to opaque; DE – 3.4; H – 6; CL – good; F – uneven to conchoidal; M – prismatic crystals, columnar aggregates, granular.
Origin and occurrence: Magmatic in alkaline syenites and their pegmatites, associated with astrophyllite, nepheline and aegirine. Crystals up to 80 mm (3 1/8 in) long come from Mount Flora, Lovozero massif, Kola Peninsula, Russia. Other localities are Narssarssuk, Greenland and Mont St.-Hilaire, Quebec, Canada.

Ajoite

$(K,Na)Cu_7AlSi_9O_{24}(OH)_6 \cdot 3\ H_2O$

TRICLINIC ●

Properties: C – blue-green; S – light green; L – vitreous to dull; D – translucent to opaque; DE – 3.0; H – not determined; CL – not determined; F – uneven to conchoidal; M – prismatic crystals, massive.
Origin and occurrence: Secondary in Cu deposits, associated with shattuckite. It occurs rarely as small crystals in the New Cornelia mine, Ajo, Arizona, USA. It is also known as inclusions in quartz crystals in the Messina mine, Transvaal, South Africa.

Ajoite, 60 mm, Ajo, U.S.A.

Holmquistite, 60 mm, Lac Malartic, Canada

serpentinites and at the contact of serpentinites and pegmatites, usually with actinolite or tremolite and cordierite. Crystals up to 150 mm (6 in) long come from the Marbridge No.1 mine, Quebec, Canada. It is also known from Bodenmais, Germany; Snarum, Sweden and Heřmanov, Czech Republic.

Antophyllite

AMPHIBOLE GROUP

$Mg_7Si_8O_{22}(OH,F)_2$

ORTHORHOMBIC ● ● ● ●

Properties: C – white, light brown, yellow, light green; S – white; L – vitreous to dull; D – transparent to translucent; DE – 3.4; H – 5.5-6; CL – good; F – uneven to conchoidal; M – long prismatic crystals, columnar and radial aggregates, granular.
Origin and occurrence: Metamorphic in gneisses,

Anthophyllite, 60 mm, North Carolina, U.S.A.

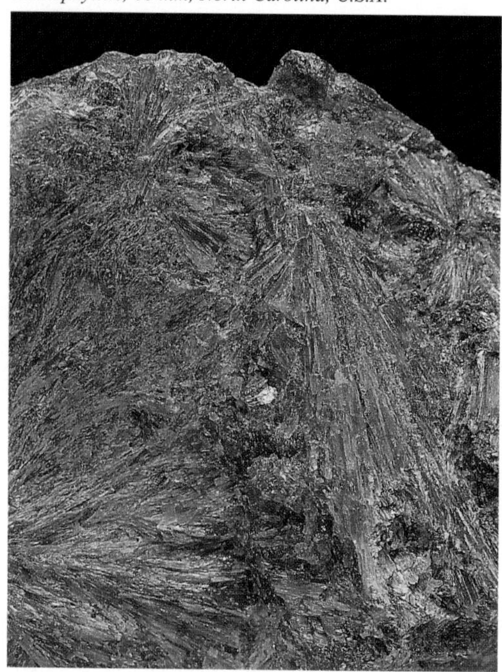

Holmquistite

AMPHIBOLE GROUP

$Li_2Mg_3Al_2Si_8O_{22}(OH,F)_2$

ORTHORHOMBIC ● ● ●

Properties: C – blue, purple, gray, black; S – white; L – vitreous to dull; D – transparent to translucent; DE – 3.0; H – 5.5-6; CL – good; F – uneven to conchoidal; M – acicular and prismatic crystals, fibrous aggregates, granular.
Origin and occurrence: Hydrothermal at the contacts between complex Li-bearing pegmatites and amphibolites, sometimes associated with biotite. It occurs in Greenbushes, Western Australia, where its fibers reach up to 180 mm (7 1/16 in) in length. It is also known from Utö, Sweden; Manono, Zaire; and Brandbrücken, Austria.

Manganogedrite

AMPHIBOLE GROUP

$Mn_2Fe_5Si_8O_{20}(OH)_2$

MONOCLINIC ● ●

Properties: C – gray, dark green, brown, greenish; S – white; L – vitreous to dull; D – translucent to opaque; DE – 3.5; H – 5-6; CL- good; F – uneven to conchoidal; M – columnar, acicular and radial aggregates, granular.
Origin and occurrence: Metamorphic in contact and regionally metamorphosed Mn rich rocks, associated with chlorite and magnetite. It comes from Dannemora, Sweden and elsewhere.

Manganogedrite, 70 mm, Franz Joseph Iceberg, New Zealand

Tremolite, 37 mm, Maricopa Co., Arizona, U.S.A.

Actinolite, 30 mm xx, Austris

Tremolite

AMPHIBOLE GROUP

$Ca_2Mg_5Si_8O_{20}(OH)_2$

MONOCLINIC ● ● ● ● ●

Properties: C – white, gray, greenish, green, brown, pink; S – white; L – vitreous to dull; D – transparent to translucent; DE – 2.9; H – 5-6; CL – good; F – uneven to conchoidal; M – columnar, acicular and radial aggregates, granular.
Origin and occurrence: Metamorphic in contact and regionally metamorphosed rocks, dolomites and ultrabasic rocks; hydrothermal in the Alpine-type veins, typically associated with diopside, talc, dolomite and calcite. It is a typical rock-forming mineral, known from many localities. Prismatic crystals up to 40 cm (15¾ in) long come from Brumado, Bahía, Brazil. It occurs also in Campolungo, Switzerland; Zillertal, Austria; Gouverneur, New York, USA and elsewhere.

Actinolite

AMPHIBOLE GROUP

$Ca_2(Mg,Fe)_5Si_8O_{20}(OH)_2$

MONOCLINIC ● ● ● ● ●

Properties: C – light green to almost black; S – white; L – vitreous to dull; D – transparent to translucent; DE – 3.2; H – 5-6; CL – good; F – uneven to conchoidal; M – columnar, acicular and radial aggregates, granular.
Origin and occurrence: Metamorphic in contact and regionally metamorphosed rocks, also in dolomites and some basic rocks (amphibolites, shales). It is a typical rock-forming mineral, associated with anthophylite, chlorite, talc, dolomite and calcite. It is known from many localities where it forms columnar aggregates up to 250 mm (9¹³/₁₆ in) long, as in Knappenwand and Zillertal, Austria; Val Malenco, Italy; Brumado, BahRa, Brazil and Sobotín, Czech Republic.

Edenite

AMPHIBOLE GROUP

$NaCa_2(Mg,Fe)_5Si_7AlO_{22}(OH)_2$

MONOCLINIC ● ● ● ●

Properties: C – white, gray to dark green; S – white; L – vitreous to dull; D – transparent, translucent to opaque; DE – 3.1; H – 5-6; CL – good; F – uneven to conchoidal; M – prismatic crystals, granular.
Origin and occurrence: Metamorphic in regionally metamorphosed basic rocks and marbles; magmatic in diorites and gabbros. Typical rock-forming mineral, its crystals up to 40 mm (1⁹/₁₆ in) long come from Edenville, New York, USA and Bancroft, Ontario, Canada.

Edenite, 48 mm, Wilberforce, Canada

Pargasite

AMPHIBOLE GROUP

$NaCa_2(Mg,Fe)_4AlSi_6Al_2O_{22}(OH)_2$

MONOCLINIC ● ● ● ●

Properties: C – light brown, green-blue, brown to black-brown; S – white; L – vitreous to dull; D – transparent to translucent; DE – 3.1; H – 5-6; CL – good; F – uneven to conchoidal; M – prismatic crystals, granular, massive.
Origin and occurrence: Metamorphic in regionally metamorphosed basic rocks, marbles and skarns; magmatic in diorites and gabbros; typical rock-forming mineral. Crystals up to 80 mm (3¹/₈ in) across come from the Jensen quarry, Riverside, California, USA; Pargas, Finland; and Hunza, Pakistan.

Common Amphibole

AMPHIBOLE GROUP

$(Ca,Na)_2(Mg,Fe,Al)_5(Si,Al)_8O_{22}(OH,F)_2$

MONOCLINIC ● ● ● ● ●

Properties: C – dark green, brown to black; S – gray; L – vitreous to dull; D – translucent to opaque; DE – 3.4; H – 5-6; CL – good; F – uneven to conchoidal; M – short prismatic crystals, columnar aggregates, granular.
Origin and occurrence: Metamorphic in basic rocks, as amphibolites and shales; magmatic in some igneous rocks, as syenites, diorites, andesites and basalts. Typical rock-forming mineral, known from many localities. Large crystals up to 1 m (39³/₈ in) long come from Silver Crater, Ontario, Canada. It is also known from Zillertal, Austria; Lukov, Czech Republic and elsewhere.

Common Amphibole, 50 mm x, Lukov, Czech Republic

Richterite

AMPHIBOLE GROUP

$Na_2Ca(Mg,Fe,Al)_5(Si,Al)_8O_{22}(OH,F)_2$

MONOCLINIC ● ● ●

Properties: C – brown, yellow, dark red, dark green; S – gray; L – vitreous to dull; D – transparent to translucent; DE – 3.1; H – 5-6; CL – good; F – uneven to conchoidal; M – long prismatic crystals, columnar aggregates, granular.
Origin and occurrence: Magmatic in alkaline rocks, as trachytes and alkaline granites; rare metamorphic in some metamorphosed rocks, as skarns and marbles. Crystals up to 100 mm (4 in) long come from Wilberforce, Ontario, Canada and Langban, Sweden.

Pargasite, 70 mm, Limberg, Parainen, Finnland

Richterite, 40 mm, Wilberforce, Canada

Riebeckite, 45 mm, Griqualand, South Africa

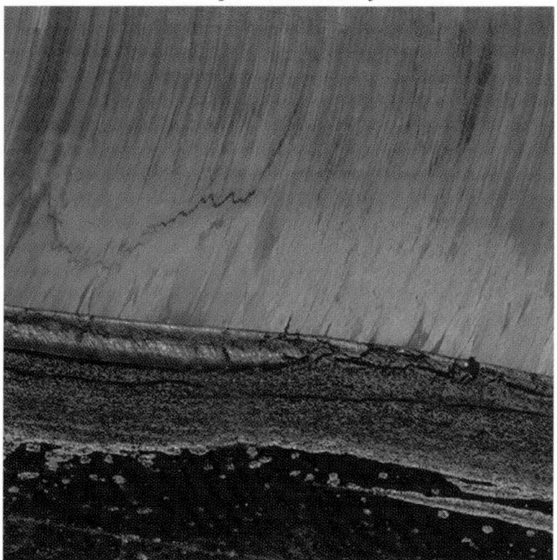

Riebeckite

AMPHIBOLE GROUP

$Na(Fe^{2+},Mg)_3Fe^{3+}{}_2Si_8O_{22}(OH)_2$

MONOCLINIC ● ● ● ●

Varieties: crocidolite (tiger's eye)

Properties: C – dark blue to black, crocydolite – gray-blue; S – white; L – vitreous to dull; D – translucent to opaque; DE – 3.4; H – 5; CL – perfect; F – uneven to conchoidal; M – long prismatic crystals, acicular to fibrous aggregates.
Origin and occurrence: Magmatic in alkaline rocks, as granites, syenites and rhyolites; metamorphic in regionally metamorphosed Fe rich shales. Large crystals up to 150 mm (6 in) long come from Khangay, Mongolia. Crocidolite is famous from Griqualand, South Africa.

Wollastonite, 44 mm, Crestmore, U.S.A.

Aenigmatite, 40 mm xx, Eveslogchorr, Kola, Russia

Aenigmatite

$Na_2Fe_5TiSi_6O_{20}$

TRICLINIC ● ●

Properties: C – black; S – red-brown; L – submetallic to dull; D -almost opaque; DE – 3.8; H – 5.5; CL – perfect; F – uneven to conchoidal; M – long prismatic crystals, columnar aggregates, granular.
Origin and occurrence: Magmatic in alkaline syenites and volcanic rocks, associated with aegirine and arfvedsonite. It occurs in Julianehab, Greenland; Lipari Island, Italy and elsewhere.

Wollastonite

$Ca_3Si_3O_9$

TRICLINIC ● ● ● ●

Properties: C – white, gray, light greenish, pink; S – white; L – vitrous to dull; D – transparent to translucent; DE – 3.0; H – 4.5-5; CL – good; F – uneven to conchoidal; M – tabular to short prismatic crystals, columnar and radial aggregates, granular; LU – orange.
Origin and occurrence: Metamorphic mainly in contact metamorphosed rocks, as marbles, skarns, less common in pyroxene gneisses and quartzites, associated with grossular, diopside and vesuvianite. Columnar and acicular aggregates up to 180 mm ($7^1/_{16}$ in) long occur in the Strickland quarry, Connecticut, USA. It also come from Crestmore, California, USA; Ciclova, Romania and Žulová, Czech Republic. Short prismatic crystals up to 30 cm (12 in) across are known from the Santa Fe mine, Chiapas, Mexico.

Bustamite, 20 mm xx, Broken Hill, Australia

Bustamite

$(Mn,Ca)_3Si_3O_9$

TRICLINIC ● ● ●

Properties: C – pink to red-brown; S – white; L – vitreous to dull; D – transparent to translucent; DE – 3.4; H – 5.5-6.5; CL – good; F – uneven to conchoidal; M – tabular crystals, acicular aggregates, massive.

Origin and occurrence: Metamorphic or hydrothermal in Mn rich metamorphic rocks or skarns. It is associated with rhodonite and other Mn silicates in Franklin, New Jersey, USA; Langban, Sweden; and Broken Hill, New South Wales, Australia.

Pectolite

$Ca_2NaSi_3O_8(OH)$

TRICLINIC ● ● ●

Properties: C – white, pinkish; S – white; L – vitreous to dull; D – transparent to translucent; DE – 2.9; H – 4.5-5; CL – good; F – uneven to conchoidal; M B acicular and radial aggregates; LU – locally yellow to orange.

Origin and occurrence: Hydrothermal in basalt cavities, less frequently along the cracks in marbles, associated with zeolites. Classic localities are West Paterson, New Jersey, USA, where it forms acicular spherical aggregates up to 180 mm ($7^1/_{16}$ in) across; also Monte Baldo, Italy; Želechovské údolí, Czech Republic. Prismatic crystals up to 50 mm (2 in) long come from Mont St.-Hilaire, Quebec, Canada.

Sérandite

$Mn_2NaSi_3O_8(OH)$

TRICLINIC ●

Properties: C – pink, red; S – white; L – vitreous to dull; D – transparent to translucent; DE – 2.9; H – 4.5-5; CL – good; F – uneven to conchoidal; M – tabular and prismatic crystals, granular.

Origin and occurrence: Hydrothermal in cavities of igneous rocks (alkaline syenite, carbonatite) together with aegirine and analcime. Perfect pink crystals up to 200 mm ($7^7/_8$ in) long come from Mont St. Hilaire,

Pectolite, 70 mm, Želechovské údolí, Czech Republic

Sérandite, 25 mm, Mont St.-Hilaire, Canada

Quebec, Canada; crystals up to 60 mm ($2^3/_8$ in) long found in the Yubileinaya dike, Mount Karnasurt, Lovozero massif, Kola Peninsula, Russia.

Charoite

$(K,Sr,Ba)(Ca,Na)_2(Si,Al)_4O_{10}(OH,F)$

TRICLINIC ● ●

Properties: C – purple; S – white; L – vitreous to dull; D – translucent to opaque; DE – 2.6; H – 5-6; CL – good; F – uneven; M – fibrous aggregates, massive.
Origin and occurrence: Hydrothermal in alkaline igneous rocks, associated with aegirine and nepheline. Rich aggregates occur in Sirenevyi Kamen, Chara river basin, Murun massif, Yakutia, Russia.
Application: cut and polished as decorative stone.

Agrellite

$NaCa_2Si_4O_{10}F$

TRICLINIC ●

Properties: C – gray-white to greenish; S – white; L – vitreous, dull to pearly; D – transparent to translucent; DE – 2.9; H – 5.5; CL – good; F – uneven; M – long prismatic crystals, granular.
Origin and occurrence: Metamorphic in alkaline gneisses. It occurs as prismatic crystals up to 100 mm (4 in) long in the Kipawa river basin, Quebec, Canada.

Okenite

$CaSi_2O_4(OH)_2 . H_2O$

TRICLINIC ● ●

Properties: C – white, yellowish; S – white; L – vitreous to dull; D – transparent to translucent; DE – 2.3; H – 4.5-5; CL – good; F – uneven to conchoidal; M B blade-shaped crystals, acicular crystals and their spherical aggregates.

Agrellite, 120 mm, Kipawa, Canada

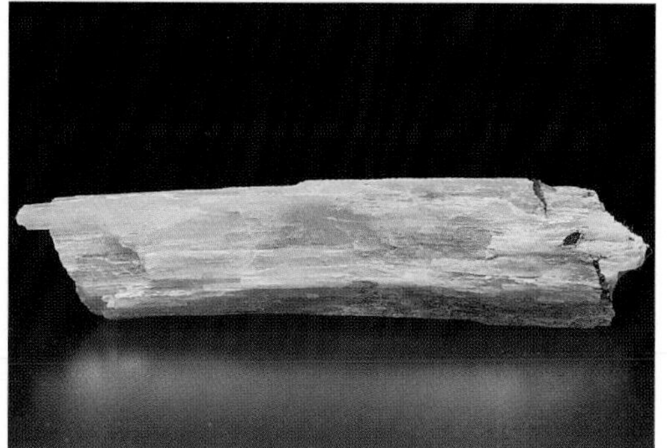

Charoite, 50 mm, Sirenevyi Kamen, Russia

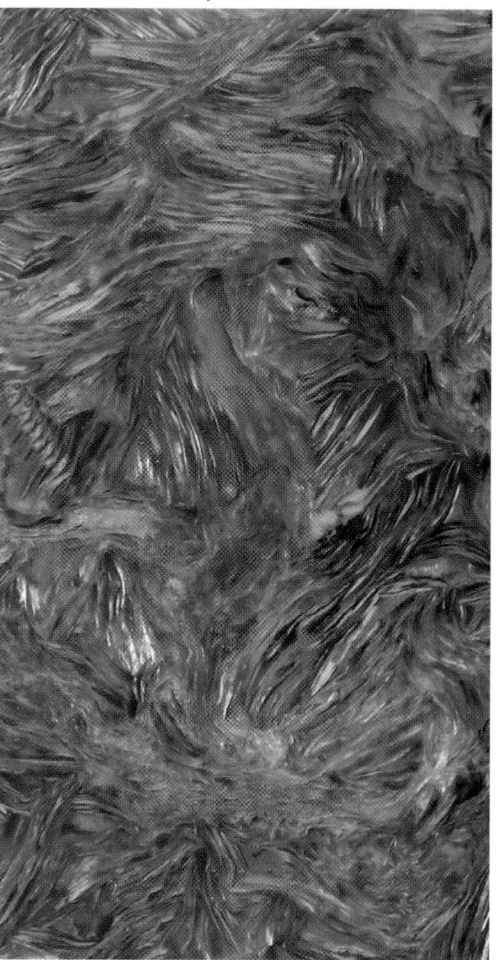

Origin and occurrence: Hydrothermal in basalt cavities together with zeolites. Rich acicular spherical aggregates up to 80 mm ($3^1/_8$ in) in diameter come from basalt cavities in the vicinity of Poona, India; also known from Faeroe Islands.

Okenite, 125 mm, Mumbai, India

Xonotlite, 20 mm vein, Staré Ransko, Czech Republic

Xonotlite

$Ca_6Si_6O_{17}(OH)_2$

MONOCLINIC ● ●

Properties: C – white, pinkish, gray; S – white; L – vitreous, dull to pearly; D – transparent to translucent; DE – 2.7; H – 6.5; CL – good; F – uneven to conchoidal; M – acicular to fibrous aggregates, massive.

Origin and occurrence: Hydrothermal in cavities of basic and ultrabasic rocks and in the Alpine-type veins, associated with calcite and zeolites; rare metamorphic in skarns and in contact metamorphosed marbles. It comes from Tetela de Xonotla, Puebla, Mexico; Crestmore, California, USA; and Mihara mine, Japan. It is was also found in Staré Ransko, Czech Republic.

Elpidite

$Na_2ZrSi_6O_{15} \cdot 3H_2O$

ORTHORHOMBIC ● ●

Properties: B – colorless, yellowish, red; S – white; L – vitreous to dull; D – transparent to translucent; DE – 2.6; H – 5.5-6.5; CL – good; F – uneven to conchoidal; M – long prismatic crystals, fibrous aggregates.

Origin and occurrence: Hydrothermal in albitized parts of alkaline pegmatites, associated with albite and aegirine. Perfect crystals up to 30 cm (12 in) long come from Tarbagatay, Kazakhstan. Crystals up to 200 mm ($7^7/_8$ in) long found at Mont St.-Hilaire, Quebec, Canada. Also known from Khan Bogdo, Gobi desert, Mongolia; Umbozero mine on Mount Alluaiv, Lovozero massif, Kola Peninsula, Russia.

Elpidite, 128 mm, Mont St.-Hilaire, Canada

Rhodonite, 35 mm x, Broken Hill, Australia

Rhodonite, 70 mm, Maloye Sedelnikovo, Ural Mts., Russia

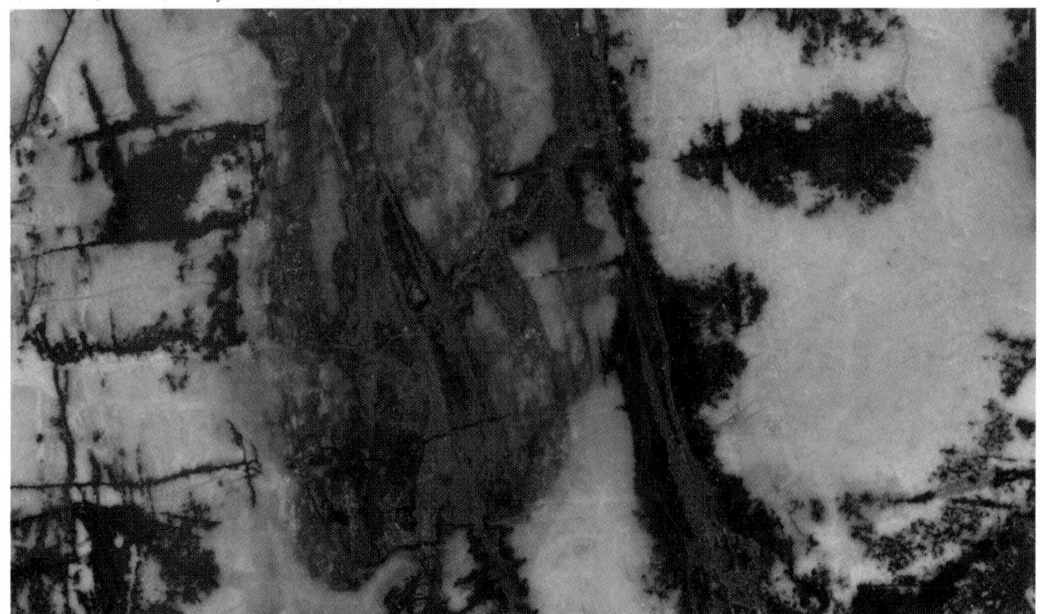

Rhodonite

$CaMn_4Si_3O_9$

TRICLINIC ● ● ●

Properties: C – pink, red, red-brown, gray; S – white; L – vitreous to dull; D – transparent to translucent; DE – 3.7; H – 5.5-6.5; CL – good; F – uneven to conchoidal; M – tabular crystals, granular, massive.
Origin and occurrence: Metamorphic in Mn-rich rocks, hydrothermal in ore veins, associated with spessartine and Mn oxides. Perfect crystals up to 200 mm (7⁷/₈ in) long come from Franklin, New Jersey, USA; also from Broken Hill, New South Wales, Australia. Granular and massive aggregates occur in Langban, Sweden and in Maloye Sidelnikovo, Ural mountains, Russia.
Application: cut and polished as a decorative stone.

Babingtonite

$Ca_2Fe^{2+}Fe^{3+}Si_5O_{14}(OH)$

TRICLINIC ● ●

Properties: C – green-black to black-brown; S – green-gray; L – vitreous to dull; D – translucent to opaque; DE – 3.3; H – 5. 5-6; CL – good; F – uneven to conchoidal; M – short prismatic crystals, granular.
Origin and occurrence: Hydrothermal along the cracks in rocks, associated with prehnite, adularia and epidote. Well-formed crystals about 20 mm (²⁵/₃₂ in) across are known from Arendal, Norway. Other localities are Westfield, Massachusetts, USA, and Baveno, Italy.

Babingtonite, 5 mm x, Mumbai, India

Kinoite

$Ca_2Cu_2Si_3O_8(OH)_4$

MONOCLINIC ●

Properties: C – blue; S – light blue; L – vitreous to dull; D – transparent to translucent; DE – 3.2; H – 5; CL B perfect; F – uneven to conchoidal; M – prismatic crystals, granular.
Origin and occurrence: Hydrothermal along cracks in rocks, associated with apophyllite and copper. Small crystals were found in the Christmas mine, Santa Rita mountains, Arizona, also in the Kearsarge vein, Keweenaw Peninsula, Michigan, USA.

Inesite

$Ca_2Mn_7Si_{10}O_{28}(OH)_2 . 5 H_2O$

TRICLINIC ● ●

Properties: C – pink, brown; S – white; L – vitreous to dull; D – translucent; DE – 3.0; H – 5.5; CL – good; F – uneven to conchoidal; M – short prismatic and tabular crystals, columnar and radial aggregates, granular.
Origin and occurrence: Hydrothermal in ore veins, associated with rhodochrosite, rhodonite and calcite. Rich aggregates and well-formed crystals about 10 mm (3/8 in) across found in Langban, Sweden; also in Hale Creek, California, USA; and Broken Hill, New South Wales, Australia. Its spherical aggregates up to 30 mm (1 3/16 in) in diameter are known from the Wessels mine, Kuruman, South Africa.

Neptunite

$KNa_2LiFe_2Ti_2Si_9O_{22}$

MONOCLINIC ● ●

Properties: C – dark brown to black; S – red-brown; L – vitreous to submetallic; D – translucent to opaque; DE – 3. 2; H – 5-6; CL – good; F – uneven to conchoidal; M – long prismatic crystals, granular.
Origin and occurrence: Hydrothermal in veins, associated with natrolite, benitoite and eudialyte. Well-formed prismatic crystals up to 80 mm (3 1/8 in) long come from the Benitoite Gem mine, San Benito Co., California, USA. It is also known from Mont St.-Hilaire, Quebec, Canada and Narssarssuk, Greenland.

Epididymite

$NaBe_2Si_3O_7(OH)$

ORTHORHOMBIC ● ●

Properties: C – white, gray, yellowish; S – white; L – vitreous to dull; D – transparent to translucent; DE – 2.6; H – 6-7; CL – good; F – uneven to conchoidal;

Kinoite, 36 mm, Christmas Mine, U.S.A.

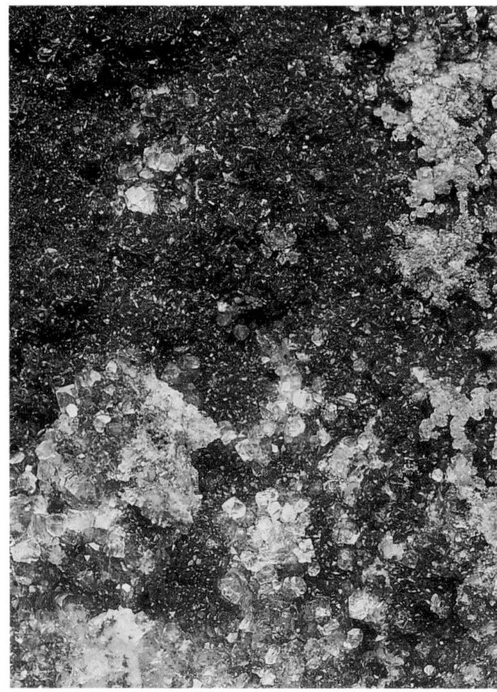

Inesite, 50 mm, Kuruman, South Africa

Neptunite, 28 mm, San Benito Co., U.S.A.

M – tabular crystals, lamellar aggregates, granular. *Origin and occurrence:* Hydrothermal in cavities of alkaline pegmatites, associated with zeolites. Tabular crystals up to 60 mm (2 3/8 in) across come from Mount Malosa, Malawi. It is also known from Langesundsfjord, Norway; Mont St.-Hilaire, Quebec, Canada; Narssarssuk, Greenland.

Bavenite, 40 mm, Yermolayevskoye, Russia

Bavenite

$Ca_4Al_2Be_2Si_9O_{26}(OH)_2$

ORTHORHOMBIC ● ● ●

Properties: C – white, yellowish, pinkish; S – white; L – vitreous to dull; D – transparent to translucent; DE – 2.8; H – 5.5; CL – good; F – uneven to conchoidal; M – prismatic to tabular crystals, radial and lamellar aggregates. *Origin and occurrence:* Hydrothermal in pegmatic cavities with albite: typically produced by beryl replacement in pegmatites or helvite replacement in skarns. Found in Baveno, Italy with crystals up to 20 mm (25/32 in) across. Also known from Strzegom, Poland. Crystals up to 4 mm (5/32 in) across found in the Hewitt quarry, USA.

Epididymite, 70 mm, Apatity, Kola, Russia

Prehnite, 28 mm, Brandberg, Namibia

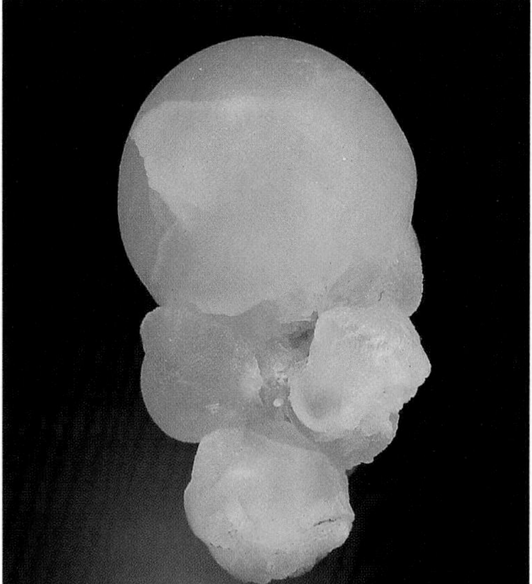

Fluorapophyllite, 47 mm, Poona, India

Prehnite

$Ca_2Al_2Si_3O_{10}(OH)_2$

ORTHORHOMBIC ● ● ● ●

Properties: C – white, greenish to green, yellow, gray; S – white; L – vitreous; PS – transparent to translucent; DE – 2.9; H – 6-6.5; CL – good; F – uneven to conchoidal; M – tabular to prismatic crystals, usually forming botryoidal or spherical aggregates, granular.
Origin and occurrence: Hydrothermal in basalt cavities and in the Alpine-type veins, commonly associated with calcite, albite, epidote and zeolites. Rich aggregates are known from West Paterson, New Jersey, USA; Val di Fassa, Italy; and Poona, India.

Astrophyllite, 30 mm xx, Khibiny Massif, Kola, Russia

Crystals up to 45 mm (1¾ in) across found in Copper Valley, Brandberg, Namibia, Asbestos, Quebec, Canada and Talnakh, Siberia, Russia.

Astrophyllite

$(K,Na)_3(Fe,Mn)_7Ti_2Si_8O_{24}(O,OH)_7$

TRICLINIC ● ●

Properties: C – various hues of yellow; S – white; L – vitreous to dull; D – transparent to translucent; DE – 3.3; H – 3; CL – good; F – uneven to conchoidal; M – long prismatic crystals, bladed aggregates.
Origin and occurrence: Magmatic in alkaline pegmatites. Crystals and their radial aggregates reaching up to 100 mm (4 in) come from Mount Eveslogchorr, Khibiny massif, Kola Peninsula., Russia. It is also known from Mont St.-Hilaire, Quebec, Canada.

Fluorapophyllite

$KCa_4Si_8O_{20}(F,OH) . 8\ H_2O$

TETRAGONAL ● ● ●

Properties: C – white, greenish yellowish, pinkish; S – white; L – greasy to pearly; D – transparent to translucent; DE – 2.4; H – 4.5-5; CL – perfect; F – uneven; M – prismatic and tabular crystals, lamellar aggregates, massive.
Origin and occurrence: Hydrothermal in cavities in volcanic rocks, along the cracks in the Alpine-type veins, in ore veins and pegmatites. Well-formed crystals up to 100 mm (4 in) long are known from Jalgaon and Poona, India; Centreville and Fairfax,

Pyrophyllite, 37 mm, California, U.S.A.

Agalmatolite, 70 mm, India

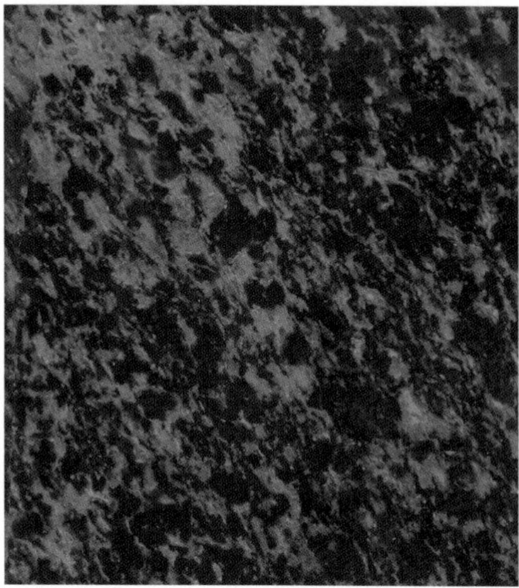

Virginia, USA. Other localities are St. Andreasberg, Germany and Bento Gonçalves, Rio Grande do Sul, Brazil.

Pyrophyllite

$Al_2Si_4O_{10}(OH)_2$

MONOCLINIC ● ● ● ●

Varieties: agalmatolite

Properties: C – white, greenish, yellow, gray; S – white; L – greasy to pearly; D – translucent to transparent; DE – 2.8; H – 1-2; CL – perfect; F – uneven; M B foliated and lamellar aggregates, massive (agalmatolite), radial to acicular aggregates.
Origin and occurrence: Metamorphic in Al-rich rocks, together with andalusite and kyanite; hydrothermal in veins with quartz and micas. Rich foliated and radial aggregates and poorly-formed crystals up to 10 mm (3/8 in) across are known from Zermatt, Switzerland; Berezovsk, Ural mountains, Russia and the Champion mine, California, USA.
Application: heat-resistant material, agalmatolite as decorative stone.

Talc

$Mg_3Si_4O_{10}(OH)_2$

MONOCLINIC ● ● ● ● ●

Properties: C – white, greenish, yellow, pinkish, gray; S – white; L – greasy to pearly; D – transparent to translucent; DE – 2.8; H – 1; CL – perfect; F – uneven; M – tabular crystals, foliated aggregates, massive.
Origin and occurrence: Metamorphic in metamorphosed basic rocks and dolomites, together with actinolite, dolomite and chlorite; hydrothermal in veins. Tabular crystals up to 20 mm (25/32 in) across come from Brumado, Bahía, Brazil and Chester, Massachusetts, USA. Rich foliated aggregates are known from Zillertal, Austria. Pseudo-morphs after quartz crystals come from St. Gotthard, Switzerland. Large deposits of massive talc are mined in China.
Application: as a filling material in textile, paper and chemical industries, heat-resistant material.

Talc, 61 mm, Rochester, U.S.A.

Muscovite

MICA GROUP

$KAl_3Si_3O_{10}(OH,F)_2$

MONOCLINIC ● ● ● ● ●

Varieties: fuchsite

Properties: C – white, greenish, yellowish, pinkish, green (fuchsite), gray; S – white; L – vitreous to pearly; D – transparent to translucent; DE – 2.8; H – 2.5-3; CL – perfect; F – uneven; M – tabular crystals, lamellar and scaly aggregates, massive.

Origin and occurrence: Metamorphic in various rock types, as mica schists, gneisses; magmatic in granites and pegmatites; hydrothermal in veins or next to ore veins (fuchsite). Important rock-forming mineral, usually associated with quartz, K-feldspar, albite and biotite. Its sheets, reached up to 5 x 3 m in size and weighed up to 85 tons in the Inikurti mine, Nellore, India. Large sheets are also known from Custer, South Dakota, USA. Well-formed crystals up to 100 mm (4 in) across come from Alabashka near Murzinka, Ural mountains, Russia. Other famous localities include Mamsk, Ural mountains, Russia and Cruzeiro mine, Minas Gerais, Brazil, where gemmy crystals were found.

Application: insulation material in electrical applications and construction.

Muscovite, 35 mm, Plumos, California, U.S.A.

Celadonite

MICA GROUP

$KFe^{3+}(Mg,Fe^{2+},Al)Si_4O_{10}(OH)_2$

MONOCLINIC ● ● ● ●

Properties: C – light green to blue-green; S – white; L – dull; D – translucent to almost opaque; DE – 3.0; H – 2; CL – perfect; F – uneven; M – earthy, sometime small scaly aggregates, massive.

Origin and occurrence: Hydrothermal as a product of mafic mineral replacement in volcanic rocks, associated with zeolites, prehnite and calcite. Famous localities are Val di Fassa and Monte Balda, Italy; also known from Bisbee, Arizona, USA.

Boromuscovite

MICA GROUP

$KAl_2BSi_3O_{10}(OH)_2$

MONOCLINIC ●

Properties: C – yellowish; S – white; L – dull; D – transparent to translucent; DE – 2.8; H – 2.5; CL – perfect; F – uneven; M – small scaly aggregates, massive.

Origin and occurrence: Hydrothermal in cavities in

Fuchsite, 4 mm xx, Pamir, Tadzhikistan

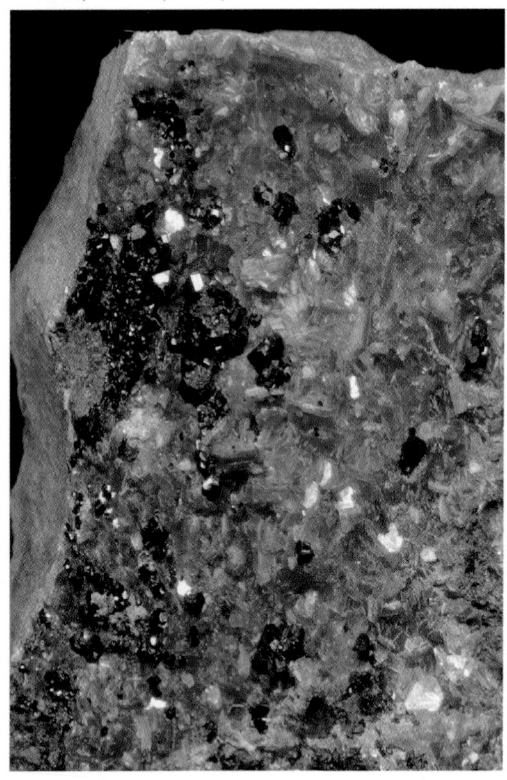

Celadonite, 60 mm, Unhošt'any, Czech Republic

complex pegmatites, associated with albite, elbaite and lepidolite. Fine-grained scaly aggregates were described from the Little Three mine, Ramona, California, USA; also in Řečice, Czech Republic.

Paragonite

MICA GROUP

$NaAl_3Si_3O_{10}(OH,F)_2$

MONOCLINIC ● ● ●

Properties: C – white, greenish, yellowish, pinkish; S – white; L – vitreous to pearly; D – transparent to translucent; DE – 2.8; H – 2.5; CL B perfect; F – uneven; M – tabular crystals, scaly aggregates, massive.
Origin and occurrence: Metamorphic in various rock types, as mica schists and gneisses. Fine-grained scaly aggregates are known from Pizzo Forno, Switzerland and elsewhere.

Boromuscovite, 40 mm, Ramona, U.S.A.

Paragonite, 80 mm, St.Gotthard, Switzerland

Glauconite

MICA GROUP

$K_{0,8}(Al,Fe^{2+},Fe^{3+},Mg)(Si,Al)_4O_{10}(OH)_2$

MONOCLINIC ● ● ● ● ●

Properties: C – light green, yellow-green to blue-green; S – light green; L – dull; PS – translucent to opaque; DE – 2.9; H – 2; CL B perfect; F – uneven; M B earthy and platy aggregates, massive.
Origin and occurrence: Hydrothermal in sedimentary and volcanic sedimentary rocks, also as a product of replacement of mafic minerals in volcanic rocks. Common in sandstones and limestones, locally associated also with phosphorites. It occurs in many localities in the Karpathians, Poland; and in the Polabí region, Czech Republic. Massive aggregates are known from the N'Chwaning No. 2 mine, Kuruman, South Africa.

Glauconite, 80 mm, Maloměřice, Czech Republic

Annite, 120 mm, Budeč, Czech Republic

Phlogopite, 45 mm, Sludyanka, Russia

Annite

MICA GROUP

$KFe_3AlSi_3O_{10}(OH,F)_2$

MONOCLINIC ● ● ●

Properties: C – black, locally with reddish or greenish tint; S – colorless; L – vitreous to pearly; D – transparent to translucent, locally opaque; DE – 2.8; H – 2.5-3; CL – perfect; F – uneven; M – tabular crystals, lamellar aggregates, massive.

Origin and occurrence: Metamorphic in skarns; magmatic in some pegmatites and granites. Its lamellar aggregates are known from Langban, Sweden and Cape Ann, Massachusetts, USA. Crystals up to 150 mm (6 in) across come from Mont St.-Hilaire, Quebec, Canada.

Phlogopite

MICA GROUP

$KMg_3AlSi_3O_{10}(OH,F)_2$

MONOCLINIC ● ● ● ● ●

Properties: C – light brown, greenish, yellowish, colorless, gray; S – white; L – vitreous to pearly; D – transparent to translucent; DE – 2.8; H – 2.5-3; CL – perfect; F – uneven; M – tabular crystals, platy and scaly aggregates, massive.

Origin and occurrence: Metamorphic in various rock types, as marbles and some ultrabasic rocks, at the contacts of pegmatites and serpentinites, magmatic in some ultrabasic rocks and pegmatites, associated with dolomite, diopside, anthophyllite. Rich scaly aggregates come from many localities, as Pargas, Finland and elsewhere. Crystals up to 5 m (16 ft) across found in Sludyanka, Siberia, Russia. Crystals up to 50 cm (20 in) across occurred in the Gardiner complex, Greenland. The largest crystals, 10 x 5 m (33 x 16 ft) across, weighing up to 90 tons, come from the Lacy mine, Ontario, Canada.

Application: as insulation material in electrical applications.

Biotite

MICA GROUP

$K(Fe,Mg)_3AlSi_3O_{10}(OH,F)_2$

MONOCLINIC ● ● ● ● ●

Properties: C – brown to black, commonly with red or green tint; S – colorless; L – vitreous to pearly; D – transparent to translucent, locally opaque; DE – 2.8; H – 2.5-3; CL – perfect; F – uneven; M – tabular crystals, scaly aggregates, massive.

Origin and occurrence: Metamorphic in various rock types, as mica schists, gneisses, migmatites and different types of metamorphosed shales; magmatic in pegmatites, granites, syenites and diorites, rare in basalts and ultrabasic rocks. Typical rock-forming mineral, usually associated with quartz, feldspars and muscovite. Rich scaly aggregates are known from a pegmatite in Evje, Norway, where its crystals reach up to several meters across. Other famous localities are Bessnes, France; Uluguru mountains, Tanzania; Silver Crater mine, Ontario, Canada; and Laacher See, Germany.

Polylithionite

MICA GROUP

$KLi_2AlSi_4O_{10}(F,OH)_2$

MONOCLINIC ● ● ●

Properties: C – gray, colorless, yellowish, purple; S – colorless; L – vitreous to pearly; D – transparent to translucent; DE – 2.8; H – 2.5-3; CL – perfect; F – uneven; M B tabular crystals, scaly aggregates, massive.

Origin and occurrence: Magmatic and locally also hydrothermal in alkaline pegmatites, granites and carbonatites, rare in Li-bearing granitic pegmatites. Well-formed crystals up to 40 mm (1⁹/₁₆ in) across are known from cavities in alkaline pegmatites from Mont St.-Hilaire, Quebec, Canada. It comes also from Illimaussaq, Greenland and from cavities in granitic pegmatites in Řečice, Czech Republic and

Biotite, 30 mm x, Ontario, Canada

elsewhere. Large industrial deposits known in Blatchford Lake, Northwest Territories, Canada.
Application: Li and Cs ore.

Trilithionite

MICA GROUP

$KLi_{1,5}Al_{2,5}Si_3O_{10}(F,OH)_2$

MONOCLINIC ● ● ● ●

Properties: C – purple, pink, blue, colorless, green; S – colorless; L – vitreous to pearly; D – transparent to translucent; DE – 2.8; H – 2.5-3; CL – perfect; F – uneven; M – tabular crystals, scaly aggregates, massive.
Origin and occurrence: Magmatic in Li-bearing pegmatites and granites, commonly associated with elbaite, spodumene, petalite, quartz and albite; sometimes hydrothermal in quartz veins and pegmatites. Well-formed crystals from pegmatite cavities are known from Virgem da Lapa, Minas Gerais, Brazil. Massive aggregates come from the Stewart Lithia mine, Pala, California and the Brown Derby No. 1 mine, Colorado, USA; Varuträsk, Sweden; Rožná, Czech Republic; and Meldon Quarry, Devon, UK.
Application: Li and Cs ore.

Polylithionite, 30 mm, Lovozero Massif, Kola, Russia

Zinnwaldite

MICA GROUP

$KLiFeAl_2Si_3O_{10}(F,OH)_2$

MONOCLINIC ● ● ●

Properties: C – gray, colorless, brown; S – colorless; L – vitreous to pearly; D – transparent to translucent; DE – 3.7; H – 2.5-3; CL – perfect; F – uneven; M – tabular crystals, scaly aggregates, massive.
Origin and occurrence: Hydrothermal in quartz veins and greisens; magmatic in pegmatites and granites, associated with fluorite, cassiterite and wolframite. Rich aggregates with scales up to 100 mm (4 in) across are known from quartz veins in Cínovec, Czech Republic. Crystals up to 150 mm (6 in) across come from pegmatite cavities in Virgem da Lapa, Minas Gerais, Brazil. It also occurs in the Pikes Peak batholith, Colorado, USA and in Baveno, Italy.
Application: Li ore.

Trilithionite, 70 mm, Tanco, Canada

Zinnwaldite, 10 mm xx, Cínovec, Czech Republic

Margarite, 73 mm, Chester, U.S.A.

Margarite

$CaAl_4Si_2O_{10}(OH)_2$

MONOCLINIC ● ● ●

Properties: C – pinkish, colorless, yellowish; S – colorless; L – vitreous to pearly; D – transparent to translucent; DE – 3.1; H – 3.5-4.5; CL B perfect; F – uneven; M – tabular crystals, platy and scaly aggregates, massive.
Origin and occurrence: Metamorphic, associated with corundum, diaspore, tourmaline and staurolite in various types of metamorphosed shales. Coarse platy aggregates come from Chester, Massachusetts and Sterling Hill, New Jersey, USA.

Clintonite

$Ca(Mg,Al)_3(Al_3Si)O_{10}(OH)_2$

MONOCLINIC ● ●

Properties: C – colorless, yellowish, pinkish, greenish; S – colorless; L – vitreous to pearly; D – transparent to translucent; DE – 3.1; H – 3.5; CL – perfect; F – uneven; M – tabular crystals, lamellar aggregates, massive.
Origin and occurrence: Metamorphic in contact metamorphosed marbles, associated with vesuvianite, grossular, diopside and spinel. Rich aggregates with lamellae up to 20 mm ($^{25}/_{32}$ in) across occur in Green Monster mountain, Alaska and Crestmore, California, USA and Monzoni, Italy.

Clintonite, 3 mm xx, Pomáz, Hungary

Stilpnomelane, 100 mm, Horní Údolí, Czech Republic

Stilpnomelane
$K(Fe^{2+},Mg,Fe^{3+})_8(Si,Al)_{12}(O,OH)_{27}$

MONOCLINIC 

Properties: C – black, black-brown, black-green, yellow-brown; S – colorless; L – vitreous to dull; D – translucent to opaque; DE – 2.8; H – 3; CL – perfect; F – uneven; M – tabular crystals, foliated, lath-like and acicular aggregates..
Origin and occurrence: Metamorphic in Fe-rich shales, usually associated with chlorite, magnetite and albite. Crystals up to 20 mm (25/32 in) across come from Jim Pond township, Maine, USA. Foliated aggregates are known from Horní Údolí near Zlaté Hory, Czech Republic and in Mesabi Range, Minnesota, USA.

Montmorillonite
$(Ca,Na)_{0,33}(Al,Mg)_2Si_4O_{10}(OH)_2 \cdot n\ H_2O$

MONOCLINIC 

Properties: C – white, yellowish, greenish, bluish; S – white; L – greasy; D – translucent to opaque; DE – 2.3; H – 1-2; CL – perfect, wet massive aggregates are plastic; F – uneven; M – earthy aggregates, massive.
Origin and occurrence: Hydrothermal as a product of replacement of other minerals in volcanic rocks, granitic pegmatites and sediments. It is abundant in many localities. Montmorillonite deposits are known from Antrim, Northern Ireland, UK; in Hungary; Slovakia and many places in the USA.
Application: ceramics and chemical industry.

Saponite
$(Ca,Na)_{0,3}(Fe,Mg)_3(Si,Al)_4O_{10}(OH)_2 \cdot 4\ H_2O$

MONOCLINIC 

Properties: C – white, yellowish, gray-green, bluish; S – white; L – greasy; D – translucent to opaque; DE – 2.1; H – 1-2; CL – perfect, massive aggregates are plastic under wet conditions; F – uneven; M – earthy, nodular, massive.
Origin and occurrence: Hydrothermal as a product of replacement of other minerals in volcanic rocks and serpentinites. It occurs together with copper in many places in Keweenaw Peninsula, Michigan, USA. Scales up to 10 mm (3/8 in) across described from Mont St.-Hilaire, Quebec, Canada.

Montmorillonite, Kisthaas, Syria

Saponite, 2 mm crust, Erdöbénye, Hungary

Vermiculite
$(Fe,Mg,Al)_3(Si,Al)_4O_{10}(OH)_2 \cdot 4\ H_2O$

MONOCLINIC 

Properties: C – yellow-brown, gray-green, green-brown; S – white; L – greasy; D – translucent to opaque; DE – 2.5; H – 1.5; CL B perfect; F – uneven; M – scaly aggregates, massive; R – it expands when heated.
Origin and occurrence: Hydrothermal as a product of replacement of phlogopite and other mafic micas in various rock types. It occurs in Palbora, South Africa; Milbury, Massachusetts, USA; and Kovdor massif, Kola Peninsula, Russia.
Application: electrical applications and paper industry.

Vermiculite, 50 mm, Drahonín, Czech Republic

Cookeite, 60 mm, Dobrá Voda, Czech Republic

Cookeite

CHLORITE GROUP

$LiAl_4Si_3O_{10}(OH)_8$

MONOCLINIC ● ●

Properties: C – white, yellowish, pinkish, brown; S – colorless; L – vitreous to pearly; D – transparent to translucent; DE – 2.6; H – 2.5-3.5; CL – perfect; F – uneven; M B scaly, locally radial aggregates, massive.

Origin and occurrence: Hydrothermal in cavities in Li-bearing pegmatites, associated with elbaite, lepidolite, fluorapatite and albite, it also replaces spodumene; metamorphic in Al-rich shales, together with diaspore and pyrophyllite. Massive aggregates, replacing spodumene, occur in the Tanco mine, Bernic Lake, Manitoba, Canada. It also comes from cavities in pegmatites in the Little Three mine, Ramona and the Himalaya mine, Mesa Grande, California. It is also known from the Pulsifer quarry, Maine, USA and Muiane, Mozambique.

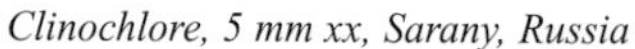

Clinochlore, 5 mm xx, Sarany, Russia

Clinochlore

CHLORITE GROUP

$(Mg,Fe)_5Al_2Si_3O_{10}(OH)_8$

MONOCLINIC ● ● ● ● ●

Varieties: kämmererite

Properties: C – greenish, gray, white, yellowish, brown, red-purple (kämmererite); S – colorless; L – vitreous to pearly; D – transparent to translucent; DE – 2.7; H – 2-2.5; CL – perfect; F – uneven; M B poorly-developed tabular crystals, foliated and radial aggregates, massive.

Kämmererite, 11 mm xx, Kop Daglari, Turkey

Chamosite, 70 mm, Hajan, Hungary

Dickite, 100 mm, Nowa Ruda, Poland

Origin and occurrence: Metamorphic in various types of shales and marbles; hydrothermal in quartz veins and Alpine-type veins, occasionally replaces certain minerals as biotite. Crystals up to 50 mm (2 in) across, associated with chondrodite and magnetite, are known from the Tilly Foster mine, Brewster, New York and Chester, Pennsylvania, USA. It also occurs in Val d'Ala, Italy and Zillertal, Austria. Scales up to 40 cm (15¾ in) across come from Beramy, Madagascar. Kämmererite crystals up to 20 mm (25/32 in) across found in Kop Daòlari, Erzerum, Turkey.

Chamosite

CHLORITE GROUP

$(Fe,Mg)_5Al_2Si_3O_{10}(OH)_8$

MONOCLINIC ● ● ● ●

Properties: C – gray, gray-green, brown; S – gray-green; L – vitreous to dull; D – translucent to opaque; DE – 3.2; H – 2-2.5; CL – perfect; F – uneven; M B scaly and oolitic aggregates, massive.
Origin and occurrence: Metamorphic to hydrothermal in various types Fe-rich sediments, typically associated with siderite and magnetite. It is common in Chamoson, Switzerland and Nucice, Czech Republic. Spherical aggregates up to 15 mm (19/32 in) in diameter found in Mont St.-Hilaire, Quebec, Canada.

Kaolinite

$Al_2Si_2O_5(OH)_4$

TRICLINIC ● ● ● ● ●

Properties: C – white, yellowish, greenish, gray; C – colorless; L – earthy; D – translucent to opaque; DE – 2.6; H – 1; CL – perfect; F – uneven; M – earthy and exceptionally scaly aggregates, massive.
Origin and occurrence: Hydrothermal, as a result of feldspar replacement in various rock types, as granites and arcoses. It forms large deposits in China, France, UK and Czech Republic.
Application: raw material for ceramics.

Dickite

$Al_2Si_2O_5(OH)_4$

MONOCLINIC ● ● ●

Properties: C – white, yellowish; S – colorless; L – earthy to pearly; D – translucent to opaque; DE – 2.6; H – 1; CL – perfect; F – uneven; M – earthy and platy aggregates, massive.
Origin and occurrence: Hydrothermal in cavities in hydrothermal veins, together with quartz, carbonates and sulfides. It occurs in Essen, Germany; Anglesey, Wales, UK; and Kladno, Czech Republic. Microscopic crystals come from Mas d'Alary, France.

Kaolinite, 50 mm, St.Austell, UK

Nacrite, 70 mm, Rochlitz, Germany

Nacrite

$Al_2Si_2O_5(OH)_4$

MONOCLINIC ● ● ●

Properties: C – white, yellowish, gray-white; S – colorless; L – earthy to pearly; D – translucent to opaque; DE – 2.6; H – 1; CL – perfect; F – uneven; M – earthy and scaly aggregates, massive.
Origin and occurrence: Hydrothermal in cavities of hydrothermal veins, associated with quartz, fluorite and topaz. It occurs in Horní Slavkov, Czech Republic; Freiberg, Germany. Pseudo-morphs after topaz come from Ouro Preto, Minas Gerais, Brazil.

Antigorite

$Mg_3Si_2O_5(OH)_4$

MONOCLINIC ● ● ● ● ●

Properties: C – white, yellowish, greenish, gray; S – colorless to gray; L – vitreous, pearly, silky, earthy; D – translucent to opaque; DE – 2.6; H – 2.5-3.5; CL – perfect; F – uneven; M B foliated, lamellar and fibrous aggregates, massive.
Origin and occurrence: Metamorphic in serpentinites, marbles and other Mg-rich rocks; hydrothermal in veins cross-secting these rocks. It is important rock-forming mineral. It occurs in Kraubat, Austria; Hrubšice, Czech Republic; Antigorio, Italy and elsewhere.

Antigorite, 110 mm, Felsöcsatár, Hungary

Amesite

$Mg_2Al(SiAl)O_5(OH)_4$

TRICLINIC ● ●

Properties: C – greenish, gray; S – colorless; L – vitreous to pearly; D – translucent; DE – 2.8; H – 2.5-3; CL – perfect; F – uneven; M B foliated to radial aggregates, massive.
Origin and occurrence: Hydrothermal in Mg-rich rocks, associated with diaspore, magnetite and chromite. It comes from Chester, Massachusetts, USA. Crystals up to 40 mm (1⁹/₁₆ in) across found in Sarany, Ural mountains, Russia.

Cronstedtite

$Fe^{2+}{}_2Fe^{3+}{}_2SiO_5(OH)_4$

MONOCLINIC ● ●

Properties: C – black, black-brown, black-green; S – dark olive-green; L – vitreous to submetallic; D – translucent to opaque; DE – 3.6; H – 3.5; CL – perfect; F – uneven; M – prismatic crystals, columnar aggregates.
Origin and occurrence: Hydrothermal in ore veins. Fan-shaped aggregates of crystals come from Příbram and Kutná Hora, Czech Republic; Ouro Preto, Minas Gerais, Brazil; Salsigue mine, Auge, France.

Amesite, 40 mm, Sarany, Ural Mts., Russia

Cronstedtite, 42 mm, Hunan, Bolivia

Chrysocolla, 50 mm, Arizona, U.S.A.

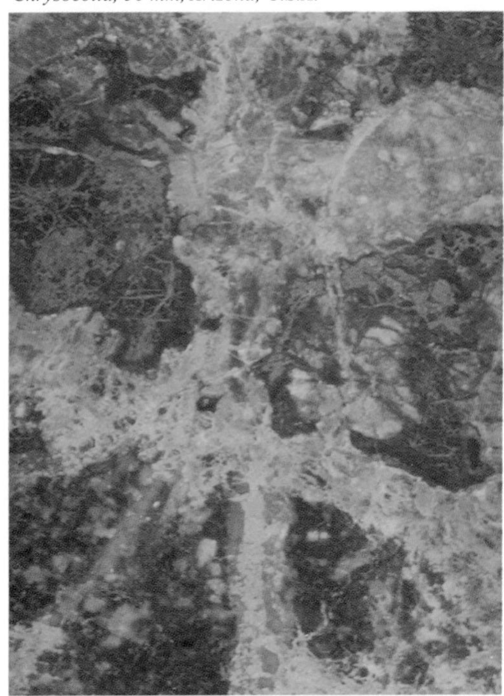

Chrysocolla

$(Cu,Al)_2H_2Si_2O_5(OH)_4 . n H_2O$

MONOCLINIC ● ● ● ●

Properties: C – different hues of blue-green; S – white; L – vitreous, greasy to earthy; D – translucent to almost opaque; DE – 2.0-2.4; H – 2-4; CL – none; F – conchoidal to uneven; M – microscopic acicular crystals, botryoidal, stalactitic and earthy aggregates. *Origin and occurrence:* Secondary in Cu deposits, associated with malachite and other secondary Cu minerals. Rich aggregates come from many places in Arizona, USA, like Bisbee and Morenci. It is also known from Mednorudnyansk, Ural mountains, Russia and Broken Hill, New South Wales, Australia.

Allophane

approximately $Al_2SiO_5 . n H_2O$

AMORPHOUS ● ● ●

Properties: C – white, gray, bluish, greenish, brown; S -white; L – vitreous, greasy and earthy; D – transparent, translucent to opaque; DE – 1.9; H – 2-3; CL – none; F – conchoidal to uneven; M – botryoidal, stalactitic and earthy aggregates.
Origin and occurrence: Hydrothermal product of alteration along the cracks of sedimentary rocks in coal deposits and in the oxidation zone of ore deposits, associated with other secondary minerals. Rich botryoidal aggregates occur in Dehr, Germany; Moldova Nuova, Romania; New Cornelia mine, Arizona, USA; and El Dragon mine, Potosi, Bolivia.

Allophane, 90 mm, Zlaté Hory, Czech Republic

Palygorskite, 120 mm, Burda, Hungary

Palygorskite

$(Mg,Al)_2Si_4O_{10}(OH) . 4 H_2O$

MONOCLINIC ● ● ● ●

Properties: C – white, gray; S – white; L – dull; D - translucent to almost opaque; DE – 2.2; H – 1; CL – good; F – uneven, aggregates plastic under wet conditions; M – acicular crystals, fibrous aggregates, massive.

Origin and occurrence: Hydrothermal as a product of alteration of different rock types, such as serpentinites, granites, marbles and graywackes, also in ore veins and the Alpine-type veins. Rich aggregates resembling leather occur along the cracks in marbles in Hejná near Horažd'ovice, Czech Republic; in the Mammoth mine, Tiger, Arizona, USA and in Palygorskaya, Russia.

Sepiolite, 70 mm, Eskisehir, Turkey

Sepiolite

$Mg_4Si_6O_{15}(OH)_2 . 6 H_2O$

ORTHORHOMBIC ● ● ●

Properties: C – white, yellowish, gray; S – white; L – dull; D – opaque; DE – 2.0; H – 2-2.5; CL – not determined; F – uneven; M – earthy aggregates, massive.

Origin and occurrence: Hydrothermal as a product of serpentinite alteration, typically associated with magnesite. Classic locality is Eskişehir, Turkey; also known from Biskoupky, Czech Republic.

Zeophylite

$Ca_4Si_3O_8(OH,F)_4 . 2 H_2O$

TRICLINIC ● ●

Properties: C – white; S – white; L – pearly; D – transparent to translucent; DE – 2.6; H – 3; CL – perfect; F – uneven; M – platy crystals, commonly with radial structure.

Origin and occurrence: Hydrothermal in cavities of volcanic rocks, associated with zeolites. Its most important locality is Radejcin, Czech Republic, where it forms spherical aggregates up to 10 mm (3/8 in) in diameter. Also known from Schellkopf, Germany and Monte Somma, Italy.

Cavansite

$CaVOSi_4O_{10} . 4 H_2O$

ORTHORHOMBIC ● ●

Properties: C – green-blue, blue; S – light blue; L – vitreous; D – transparent; DE – 2.2; H – 3-4; CL – good; F – uneven; M – long prismatic to acicular crystals, radial aggregates.

Origin and occurrence: Hydrothermal in cavities of volcanic rocks, associated with calcite, apophyllite and zeolites. Rich radial aggregates up to 30 mm

Zeophyllite, 60 mm, Radejcín, Czech Republic

Cavansite, 120 mm, Poona, India

($1^{3}/_{16}$ in) in diameter come from the Wagholi Quarry, Poona, India.

Nepheline

(K,Na)AlSiO$_4$

HEXAGONAL ● ● ● ●

Properties: C – colorless, white, greenish, yellowish, gray, green, brown; S – white; L – vitreous to greasy; D – transparent to translucent; DE – 2.7; H – 5.5-6; CL – none; F – uneven to conchoidal; M – prismatic crystals, granular, massive.
Origin and occurrence: Magmatic in many alkaline rocks, as alkaline syenites and their pegmatites, also in some basalts; rare metamorphic in gneisses. It is a typical rock-forming mineral, associated with leucite, augite and apatite. Well-formed crystals found in Mount Vesuvius, Italy. Rich aggregates are known from many localities in Khibiny massif, Kola Peninsula, Russia. Crystals up to 70 cm ($27^{9}/_{16}$ in) long occur in Davis Hill, Bancroft, Ontario. Perfect crystals up to 35 mm ($1^{3}/_{8}$ in) long, come from Mont St.-Hilaire, Quebec, Canada.
Application: locally as Al ore and in ceramic industry.

Nepheline, 43 mm, Aouli, Morocco

Petalite, 23 mm, Paprok, Afghanistan

Petalite

$LiAlSi_4O_{10}$

MONOCLINIC ● ● ●

Properties: C – colorless, white, greenish, yellowish; S – white; L – vitreous; D – transparent to translucent; DE – 2.5; H – 6-6.5; CL – good; F – uneven to conchoidal; M – prismatic crystals, granular.
Origin and occurrence: Magmatic in Li-bearing pegmatites, associated with lepidolite, elbaite, amblygonite and locally replaced by a mixture of quartz and spodumene. Gigantic crystals several meters long, come from Bikita, Zimbabwe; and Varuträsk, Sweden. Clear gemmy crystals up to 230 mm ($9^1/_{16}$ in) across found in pegmatite cavities in Paprok, Afghanistanand also in San Piero in Campo, Elba, Italy. Crystals up to 100 mm (4 in) across occur in Araçuai, Minas Gerais, Brazil.
Application: ceramic industry.

Chkalovite

$Na_2BeSi_2O_6$

ORTHORHOMBIC ● ●

Properties: C – colorless, white; S – white; L – vitreous; D – transparent to translucent; DE – 2.7; H – 6; CL – imperfect; F – conchoidal; M – prismatic crystals, granular, massive.
Origin and occurrence: Hydrothermal in alkaline pegmatites, usually associated with natrolite, sodalite and eudialyte. Crystals up to 200 mm ($7^7/_8$ in) across come from the Umbozero mine, Mount Alluaiv, Lovozero massif, Kola Peninsula, Russia; also at Julianhab, Greenland and Mont St.-Hilaire, Quebec, Canada.

Chkalovite, 70 mm, Umbozero, Kola, Russia

Sanidine, 30 mm x, Drachenfels, Germany

Sanidine
FELDSPAR GROUP
$KAlSi_3O_8$

MONOCLINIC ● ● ● ●

Properties: C – colorless, white, yellowish, pinkish, gray; S – white; L – vitreous; D – transparent to translucent; DE – 2.6; H – 6-6.5; CL – good; F – uneven to conchoidal; M – prismatic crystals, granular, massive.
Origin and occurrence: Magmatic in acid volcanic rocks, as rhyolites and trachytes, a typical rock-forming mineral. Well-formed crystals and their combinations up to 100 mm (4 in) in size occur in rhyolites and trachytes in Drachenfels and Laacher See, Germany. It also comes from Roc de Courlande, France; Beaverdell, British Columbia, Canada; and Kyustendil, Bulgaria.

Orthoclase
FELDSPAR GROUP
$KAlSi_3O_8$

MONOCLINIC ● ● ● ● ●

Varieties: adularia, moonstone (gemmy variety with chatoyancy)

Properties: C – colorless, white, yellowish, pinkish, gray, brown, yellow; S – white; L – vitreous to pearly (adularia); D – transparent to translucent; DE – 2.5; H – 6-6.5; CL – good; F – uneven to conchoidal; M – prismatic and tabular crystals and their combinations, granular, massive.
Origin and occurrence: Magmatic in rhyolites, trachytes, granites, syenites and pegmatites; metamorphic in various rock types, as orthogneisses and migmatites; hydrothermal in the Alpine-type veins, ore veins and some sediments, also in placers, a typical rock-forming mineral. Well-formed crystals and twins up to 200 mm (7⁷/₈ in) across come from granites in Twentynine Palms, California, USA; Marina di Campo, Elba, Italy; Loket and Karlovy Vary, Czech Republic; in pegmatite cavities in Strzegom, Poland; and San Piero in Campo, Elba, Italy. Yellow gemmy crystals up to 70 mm (2¾ in) across found in Itrongay, Madagascar. Gigantic feldspar crystals up to tens of meters long known from several pegmatite localities in Black Hills, South Dakota, USA; also Hagendorf, Germany. Adularia is known from Alpine–type veins in St. Gotthard, Switzerland; pebbles of moonstone occur in gem-bearing gravels in Ratnapura, Sri Lanka.
Application: ceramic and glass industry, moonstone as a gemstone.

Orthoclase, 32 mm x, Oro Grande, U.S.A.

Adularia, 60 mm, Alps, Switzerland

Microcline, 110 mm, Lake George, Teller Co., U.S.A.

Microcline

FELDSPAR GROUP

$KAlSi_3O_8$

TRICLINIC ●●●●●

Varieties: amazonite

Properties: C – colorless, white, yellowish, pinkish, gray, light to dark green (amazonite); S – white; L – vitreous; D – transparent to translucent; DE – 2.6; H – 6-6.5; CL – good; F – uneven to conchoidal; M – prismatic crystals, granular, massive.

Origin and occurrence: Magmatic in granites, syenites and pegmatites; metamorphic in various rock types, as orthogneisses and migmatites; hydrothermal in the Alpine-type veins and ore veins; a typical rock-forming mineral. Well-formed amazonite crystals up to 40 cm (15¾ in) across occur in pegmatite cavities in Crystal Peak, Colorado, USA; Keivy, Kola Peninsula, Russia; and Morefield mine, Virginia, USA. Gigantic microcline crystals up to 12 m (39 ft) across come from pegmatites in Black Hills, South Dakota, USA; Kaatiala, Finland.

Application: ceramic industry, amazonite as a decorative stone.

Hyalophane

FELDSPAR GROUP

$(K,Ba)Al(Si,Al)_3O_8$

MONOCLINIC ●●●

Properties: C – colorless, white, yellowish; S – white; L – vitreous; D – transparent to translucent; DE – 2.9; H – 6-6.5; CL – good; F – uneven to conchoidal; M – prismatic crystals, granular, massive.

Amazonite, 60 mm, Keivy, Kola, Russia

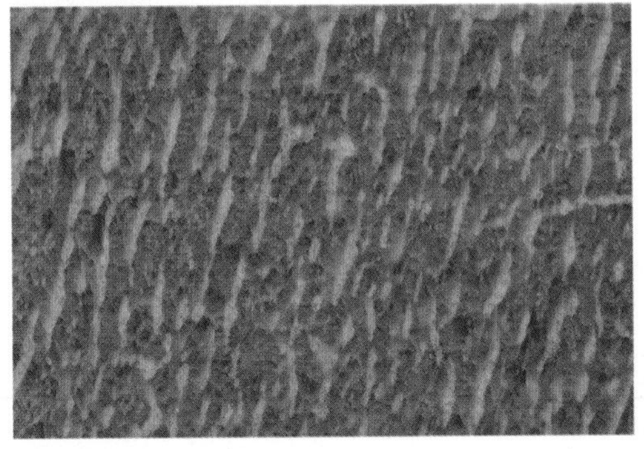

Hyalophane, 50 mm, Busovaca, Bosnia

Oligoclase, 40 mm, Ural Mts., Russia

Origin and occurrence: Magmatic in phonolites; metamorphic in various rock types, as marbles, gneisses and Mn-rich rocks; hydrothermal in the Alpine-type and ore veins. Crystals up to 100 mm (4 in) across found in Alpine-type veins near Busovaca, Bosnia-Hercegovina.

Albite

FELDSPAR GROUP

$NaAlSi_3O_8$

TRICLINIC ● ● ● ● ●

Varieties: pericline, cleavelandite

Properties: C – colorless, white, yellowish, pinkish, gray, greenish, bluish; S – white; L – vitreous; D – transparent to translucent; DE – 2.6; H – 6-6.5; CL – good; F – uneven to conchoidal; M – tabular crystals and their twins, platy aggregates, granular, massive.

Origin and occurrence: Magmatic in granites, syenites and their pegmatites; metamorphic in various rock types, as orthogneisses, migmatites, phyllites and metamorphosed shales; hydrothermal in the Alpine-type veins and ore veins, a typical rock-forming mineral. Well-formed tabular to platy cleavelandite crystals up to 150 mm (6 in) across occur in pegmatite cavities in the Amelia district, Virginia and the Pala district, California, USA. Albite crystals also come from Strzegom, Poland; San Piero in Campo, Elba, Italy; Murzinka, Ural mountains, Russia; and many localities in Minas Gerais, Brazil. Pericline crystals are known from the cracks along Alpine-type veins in Grossgreiner, Austria; St. Gotthard, Switzerland; also from ore veins in Rožňava, Slovakia.

Albite, 85 mm, Governador Valadares, Brazil

Oligoclase

FELDSPAR GROUP

$(Na_{0,9-0,7}Ca_{0,1-0,3}Al_{1,1-1,3}Si_{2,9-2,7}O_8$

TRICLINIC ● ● ● ● ●

Varieties: peristerite, sunstone

Properties: C – white, yellowish, pinkish, greenish, iridescent (peristerite); S – white; L – vitreous; D – transparent to translucent; DE – 2.6; H – 6-6.5; CL – good; F – uneven to conchoidal; M – tabular crystals, platy aggregates, granular, massive.

Origin and occurrence: Magmatic in granites, syenites, andesites and pegmatites; metamorphic in gneisses and migmatites; hydrothermal in the Alpine-type veins; typical rock-forming mineral. Large crystals, several dm long, are known mainly from pegmatites. Peristerite comes from Arendal, Norway; Miass, Ural mountains, Russia; Quadeville, Ontario, Canada; Tvedestrand, Norway. It also occurs in Vezná, Czech Republic. Sunstone is mined in the Ponderosa mine near Lakeview, Oregon, USA.

Application: peristerite and sunstone are used as gemstones.

Andesine, 50 mm, Bodenmais, Germany

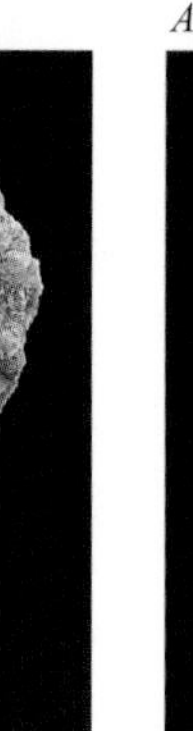

Andesine

FELDSPAR GROUP

$(Na_{0,7-0,5}Ca_{0,3-0,5}Al_{1,3-1,5}Si_{2,7-2,5}O_8$

TRICLINIC ● ● ● ●

Properties: C – yellowish, pinkish, gray, light green; S – white; L – vitreous; D – transparent to translucent; DE – 2.7; H – 6-6.5; CL – good; F – uneven to conchoidal; M – platy aggregates, granular, massive.

Labradorite, 50 mm, Labrador, Canada

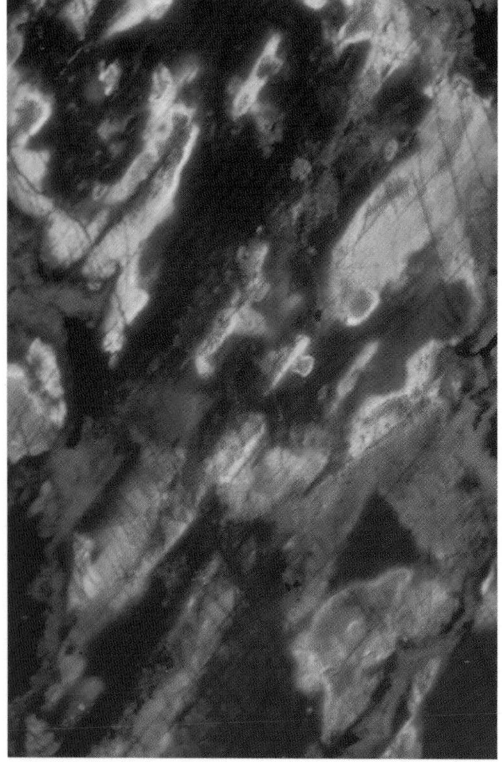

Anorthite, 10 mm xx, Pasmaada, Finnland

Origin and occurrence: C – white to dark gray, greenish; S – white; L – vitreous, pearly; D – transparent to translucent; DE – 2.7; H – 6-6.5; CL – good; F – uneven to conchoidal; M B cleavable aggregates, granular, massive.

Labradorite

FELDSPAR GROUP

$Ca_{0,7-0,5}Na_{0,3-0,5}Al_{1,5-1,7}Si_{2,5-2,3}O_8$

TRICLINIC ● ● ● ●

Properties: Magmatic in andesites, diorites; metamorphic in gneisses and migmatites; typical rock-forming mineral. Large grains are known from migmatites in Bodenmais, Germany and Adamello, Italy.
Origin and occurrence: Magmatic in gabbros, basalts, anorthosites; metamorphic in amphibolites. Large iridescent aggregates up to 1 m (39³/₈ in) across, come from Nain, Labrador, Quebec, Canada and also occur in Korostenskiy massif, Ukraine and Ylämaa, Finland.
Application: as a decorative stone.

Anorthite

FELDSPAR GROUP

$CaAl_2Si_2O_8$

TRICLINIC ● ● ●

Properties: C – gray, greenish, pinkish; S – white; L – vitreous, dull; D – transparent to translucent; DE – 2.8; H – 6-6.5; CL – good; F – uneven to conchoidal; M B granular, massive.
Origin and occurrence: Magmatic in gabbros, basalts, anorthosites; metamorphic in contact metamorphosed rocks. Pinkish grains and poorly-developed crystals come from Val di Fassa, Italy. Crystals up to 50 mm (2 in) across are known from Miyake-Jima Island, Japan. It was also found in Mount Erebus, Antarctica and Monte Somma, Italy.

Danburite

$CaB_2Si_2O_8$

ORTHORHOMBIC ● ● ●

Properties: C – colorless, white, gray, greenish, pinkish, yellow, brown, red-brown; S – white; L – vitreous, dull; D – transparent to translucent; DE – 3.0; H – 7; CL – imperfect; F – uneven to conchoidal; M – prismatic crystals, columnar aggregates, granular, massive.

Origin and occurrence: Hydrothermal in pegmatite cavities, ore veins and the Alpine-type veins; metamorphic in skarns and contact metamorphosed rocks. Well-formed prismatic crystals up to 250 mm (9 13/16 in) long come from Russell, New York, USA; Toroku, Japan; Dalnegorsk, Russia; and Charcas, San Luis Potosi, Mexico.

Danburite, 109 mm, Charcas, Mexico

Cancrinite

$Na_6CaAl_6Si_6O_{24}(CO_3) . 2 H_2O$

HEXAGONAL ● ● ●

Properties: C – colorless, white, yellow, orange, bluish; S – white; L – vitreous; D – transparent to translucent; DE – 2.4; H – 5-6; CL – good; F – uneven to conchoidal; M – prismatic crystals, granular, massive.

Origin and occurrence: Magmatic in alkaline syenites; hydrothermal as a product of replacement of volcanic rocks, associated with nepheline and sodalite. Crystals up to 20 mm (25/32 in) across come from Mont St.-Hilaire, Quebec, Canada. It was also found in Litchfield, Maine, USA; Cancrinite Hill, Bancroft, Ontario, Canada and elsewhere.

Leifite

$Na_2(Si,Al,Be)_7 (O,OH,F)_{14}$

TRIGONAL ●

Properties: C – colorless; S – white; L – vitreous; D – translucent; DE – 2.5; H – 6; CL – good; F – uneven; M – acicular crystals.

Origin and occurrence: Hydrothermal in cavities of alkaline pegmatites, associated with microcline and zinnwaldite. Crystals up to 20 mm (25/32 in) across occur in Narssarssuk, Greenland. Fine crystals are also known from Mont St.-Hilaire, Quebec, Canada, associated with serandite.

Cancrinite, 70 mm, Ditrau, Romania

Leifite, 100 mm, Mont St.-Hilaire, Canada

Sodalite, 50 mm, Bancroft, Canada

Sodalite

$Na_8Al_6Si_6O_{24}Cl_2$

CUBIC ● ● ●

Varieties: hackmanite

Properties: C – colorless, white, blue, yellow, pink (hackmanite); S – white; L – vitreous; D – transparent to translucent; DE – 2.3; H – 5.5-6; CL – imperfect; F – uneven to conchoidal; M – isometric crystals, granular, massive; LU – orange-red.
Origin and occurrence: Magmatic in alkaline syenites and phonolites, associated with nepheline, zircon and titanite; hydrothermal in marbles. Granular aggregates are known from the Princess Sodalite mine, Bancroft, Ontario, Canada and Ditrau, Romania. Crystals up to 100 mm (4 in) across come from Kangerdluarssuk, Greenland.
Application: cut as a gemstone.

Noseane

$Na_8Al_6Si_6O_{24}(SO_4)$

CUBIC ● ●

Properties: C – colorless, white, blue, gray, black; S – white; L – vitreous; D – transparent to translucent; DE – 2.3; H – 5-6; CL – good; F – uneven to conchoidal; M – isometric crystals, granular, massive.
Origin and occurrence: Magmatic in alkaline basalts and similar effusive rocks, associated with nepheline and haüyne. Granular aggregates are known from Laacher See, Germany and Monte Somma, Italy.

Lazurite

$(Na,Ca)_8Al_6Si_6O_{24}(S,SO_4)$

CUBIC ● ● ●

Properties: C – dark blue; S -light blue; L – vitreous; D – transparent to translucent; DE – 2.4; H – 5-5.5; CL – imperfect; F – uneven to conchoidal; M – isometric crystals, granular, massive.
Origin and occurrence: Metamorphic in contact metamorphosed marbles, associated with pyrite. Crystals up to 50 mm (2 in) across come from Sar-e-Sang, Badakhshan province, Afghanistan. Granular aggregates are known from many localities, such as Malobystrinskoye deposit near Lake Baikal, Russia or Monte Somma, Italy.
Application: decorative stone and gemstone.

Noseane, 6 mm grains, Laacher See, Germany

Lazurite, 60 mm, Afghanistan

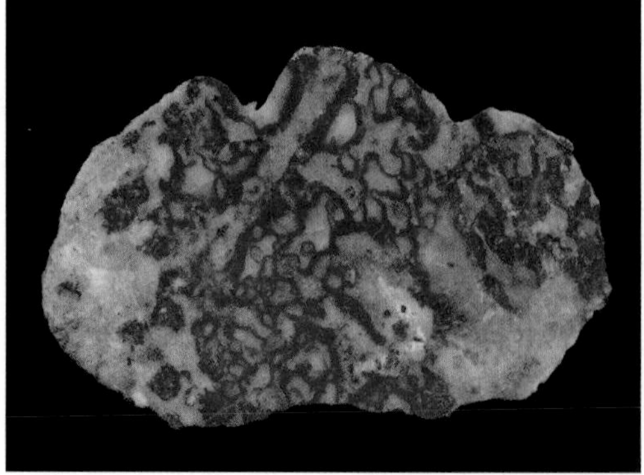

Lazurite, 21 mm x, Nuristan, Afghanistan

Lazurite, 60 mm, Chile

and IlRmaussaq, Greenland. Small crystals occur in the Poudrette quarry, Quebec, Canada.

Tugtupite

$Na_8Be_2Al_2Si_8O_{24}Cl_2$

TETRAGONAL ● ●

Properties: C – white, pink, red, bluish; S – white; L – vitreous; D – transparent to translucent; DE – 2.3; H – 5; CL – good; F – uneven to conchoidal; M – small crystals, granular, massive.
Origin and occurrence: Hydrothermal in alkaline syenites and their pegmatites, also as a product of chkalovite alteration. Aggregates up to 60 mm (2³/₈ in) across come from the Umbozero mine, Mount Alluaiv, Lovozero massif, Kola Peninsula, Russia

Tugtupite, 90 mm, Illímaussaq, Greenland

Danalite

$Fe_4Be_3Si_3O_{12}S$

CUBIC ● ●

Properties: C – gray,yellow,pink, red, brown; S – white; L – vitreous; D – transparent to translucent; DE – 3.4; H – 5.5-6; CL – imperfect; F – uneven to conchoidal; M – isometric crystals, granular, massive.
Origin and occurrence: Magmatic in granitic pegmatites; hydrothermal in greisens, skarns and ore veins. It occurs in Cape Ann, Massachusetts, USA; Hortekollen, Norway; and Coolgardie, Western Australia.

Danalite, 70 mm, Yxjöberg, Sweden

Helvite, 2 mm xx, Cavnic, Romania

Helvite

$Mn_4Be_3Si_3O_{12}S$

CUBIC ● ● ●

Properties: C – brown, gray, yellow, yellow-green; S – white; L – vitreous; D – transparent to translucent; DE – 3.4; H – 6; CL – good; F – uneven to conchoidal; M – isometric crystals, granular, massive.
Origin and occurrence: Magmatic in granitic pegmatites and alkaline syenites; hydrothermal in greisens, skarns and ore deposits. Cubic crystals up to 25 mm (1 in) across known from the Sawtooth Batholith, Idaho, USA. Crystals also come from Schwarzenberg, Germany; Cavnic, Romania; and Oslofjord, Norway.
Application: Be ore.

Scapolite, 44 mm x, Leslie Lake, Québec, Canada

Marialite

SCAPOLITE GROUP

$Na_8(AlSi_3O_8)_6\ (Cl_2,SO_4)$

TETRAGONAL ● ● ●

Properties: : C – colorless, white, gray, purplish, yellow; S – white; L – vitreous, locally pearly; D – transparent to translucent; DE – 2.5; H – 5-6; CL – good; F – uneven to conchoidal; M – prismatic crystals, columnar aggregates, granular, massive; LU – yellow to orange.
Origin and occurrence: Hydrothermal in veins, cross-cutting alkaline metamorphic rocks and in pegmatites, cross-cutting ultrabasic rocks; metamorphic in marbles and metaevaporites. It comes from Ankazobé, Madagascar; Umba, Tanzania and Crestmore, California, USA.

Meionite

SCAPOLITE GROUP

$Ca_8(Al_2Si_2O_8)_6\ (CO_3,SO_4)$

TETRAGONAL ● ● ● ●

Properties: C – colorless, white, gray, purplish, green, blue; S – white; L – vitreous; D – transparent to translucent; DE – 2.8; H – 5-6; CL – good; F – uneven to conchoidal; M – long prismatic crystals, columnar aggregates, granular, massive; LU – yellow to orange.

Leucite, 30 mm, Mt.Vesuvius, Italy

Analcime, 45 mm, Tunguzka, Russia

Origin and occurrence: Metamorphic in skarns, marbles, granulites and in contacts of volcanic rocks; hydrothermal in veins, cross-cutting Ca-rich rocks. Well formed crystals up to 40 cm (15¾ in) across come from Lake Clear and Eganville, Ontario, Canada. Crystals are also known from Monte Somma, Italy; Sludyanka, Siberia, Russia; and Pargas, Finland.

Leucite

ZEOLITE GROUP

$KAlSi_2O_6$

TETRAGONAL ● ● ● ●

Properties: C – colorless, white, gray; S – white; L – vitreous; D – transparent to translucent; DE – 2.5; H – 5.5-6; CL- imperfect; F – uneven to conchoidal; M – isometric crystals, granular, massive.
Origin and occurrence: Magmatic in effusive K-rich basalts, associated with nepheline and sanidine. Well-formed crystals, several cm in size, come from Mount Vesuvius, Italy and Laacher See, Germany.
Application: ceramic industry.

Analcime

ZEOLITE GROUP

$NaAlSi_2O_6 . H_2O$

CUBIC ● ● ● ●

Properties: C – colorless, white, pinkish, yellowish; S – white; L – vitreous; D – transparent to translucent; DE – 2.3; H – 5-5.5; CL – imperfect; F – uneven to conchoidal; M – isometric crystals, granular, massive.
Origin and occurrence: Mainly hydrothermal as a product of nepheline or sodalite replacement; rare magmatic in effusive rocks; also in sediments, associated with calcite and zeolites. Well-formed crystals up to 30 cm (12 in) across come from Nidym, Siberia, Russia; Lago Maggiore, Italy; Mont St.-Hilaire, Quebec, Canada; West Paterson, New Jersey, USA and elsewhere.

Pollucite

ZEOLITE GROUP

$(Cs,Na)AlSi_2O_6 . H_2O$

CUBIC ● ● ●

Properties: C – colorless, white, gray; S – white; L – vitreous; D – transparent to translucent; DE – 2.9; H – 6.5-7; CL – none; F – uneven to conchoidal; M – isometric crystals, granular, massive.
Origin and occurrence: Magmatic and rarely also hydrothermal in Li-bearing pegmatites, associated with lepidolite, albite, quartz and petalite. Almost monomineral layer of pollucite several meters thick occurs in the Tanco mine, Bernic Lake, Manitoba, Canada. White and colorless crystals from pegmatite cavities up to 60 cm (24 in) across come from Paprok, Afghanistan. Also known from San Piero in Campo, Elba, Italy and Gilgit, Pakistan.
Application: ceramic industry, Cs ore.

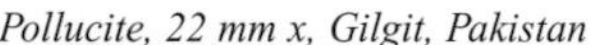

Pollucite, 22 mm x, Gilgit, Pakistan

Natrolite, 47 mm, Maricopa Co., Arizona, U.S.A.

Scolecite, 165 mm, Nasik, India

Natrolite

ZEOLITE GROUP

$Na_2Al_2Si_3O_{10} \cdot 2\,H_2O$

ORTHORHOMBIC ● ● ● ●

Properties: C – colorless, white, yellowish, pinkish; S – white; L – vitreous to silky; D – transparent to translucent; DE – 2.2; H – 5-5.5; CL – perfect; F – uneven to conchoidal; M – long prismatic to acicular crystals, fibrous and radial aggregates, granular, massive.

Origin and occurrence: Hydrothermal in cavities in volcanic rocks, alkaline pegmatites, along the Alpine-type fissures, also as a product of plagioclase replacement, commonly associated with calcite and zeolites. Colorless and white acicular crystals up to 30 cm (12 in) long come from cavities of alkaline pegmatites in Mount Putelichorr, Khibiny massif, Kola Peninsula, Russia. Crystals were also found in Narssarssuk, Greensland and Mont St.-Hilaire, Quebec, Canada. Crystals also occur in cavities of basaltic rocks in Teigarhorn, Iceland; Zálezly and Soutisky, Czech Republic; and in Faeroe Islands.

Scolecite

ZEOLITE GROUP

$CaAl_2Si_3O_{10} \cdot 3\ H_2O$

MONOCLINIC ● ● ●

Properties: C – colorless, white; S – white; L – vitreous to silky; D – transparent to translucent; DE – 2.3; H – 5; CL – good; F – uneven; M – long prismatic to acicular crystals, fibrous and radial aggregates.
Origin and occurrence: Hydrothermal in cavities in volcanic rocks and along the Alpine-type fissures, associated with calcite and zeolites. Clear prismatic crystals up to 200 mm (7 7/8 in) long come from basalt cavities near Nasik, India. Crystals come also from Teigarhorn, Iceland and Suderoy, Faeroe Islands.

Mesolite, 150 mm, Mahrashitra, India

Mesolite

ZEOLITE GROUP

$Na_{16}Ca_{16}Al_{48}Si_{72}O_{240} \cdot 64\ H_2O$

ORTHORHOMBIC ● ● ●

Properties: C – colorless, white; S – white; L – vitreous to silky; D – transparent to translucent; DE – 2.3; H – 5; CL – good; F – uneven; M – acicular crystals, fibrous and radial aggregates, granular, massive.
Origin and occurrence: Hydrothermal in cavities in volcanic rocks, associated with other zeolites. Colorless needles up to 150 mm (6 in) long come from the basalt cavities in Berufjord and Teigarhorn, Iceland. Radial aggregates up to 200 mm (7 7/8 in) in diameter known from the vicinity of Poona, India. Prismatic crystals up to 100 mm (4 in) long found in Skookumchuck Dam, Washington, USA.

Thomsonite

ZEOLITE GROUP

$Ca_2NaAl_4(Al,Si)_2Si_4O_{20} \cdot 6\ H_2O$

ORTHORHOMBIC ● ● ●

Properties: C – colorless, white, yellowish, brown-red; S – white; L – vitreous to pearly; D – transparent to translucent; DE – 2.3; H – 5-5.5; CL – good; F – uneven to conchoidal; M – prismatic crystals in clusters with radial structure, botryoidal aggregates, massive.
Origin and occurrence: Hydrothermal in cavities in volcanic rocks, also as a product of hydrothermal feldspar replacement. Colorless and white acicular crystals, forming radial aggregates up to 50 mm (2 in) in diameter come from Old Kilpatrick, Scotland, UK and West Paterson, New Jersey, USA. Hemispherical aggregates up to 30 mm (1 3/16 in) across known from Vinarická hora near Kladno, Czech Republic.

Thomsonite, 50 mm, Bazsi, Hungary

Gonnardite, 10 mm xx, Zalahaláp, Hungary

Gonnardite

ZEOLITE GROUP

$(Na,Ca)_{6-8}(Al,Si)_{20}O_{40} \cdot 12\ H_2O$

TETRAGONAL ● ● ●

Properties: C – colorless, white, yellowish; S – white; L – vitreous to pearly; D – transparent to translucent; DE – 2.3; H – 4.5-5; CL – good; F – uneven to conchoidal; M – prismatic crystals, fibrous and radial aggregates.
Origin and occurrence: Hydrothermal in cavities in volcanic rocks, associated with zeolites, also in the contact zone of marbles, associated with wollastonite. Acicular crystals and fibrous aggregates occur in Aci Castello, Sicily, Italy; Weilberg, Germany; and Bundoora, Victoria, Australia.

Edingtonite

ZEOLITE GROUP

$BaAl_2Si_3O_{10} \cdot 3\ H_2O$

TETRAGONAL ● ● ●

Properties: C – white, gray, pinkish; S – white; L – vitreous; D – transparent to translucent; DE – 2.8; H – 4; CL – good; F – uneven to conchoidal; M – prismatic crystals, granular, massive.
Origin and occurrence: Hydrothermal in cavities and along the cracks in volcanic rocks, associated with calcite and harmotome. Colorless and white prismatic crystals up to 40 mm ($1^{9}/_{16}$ in) across come from Old Kilpatrick, Scotland, UK; Staré Ransko,

Edingtonite, 100 mm, Staré Ransko, Czech Republic

Czech Republic; and Mont St.-Hilaire, Quebec, Canada.

Dachiardite-Ca

ZEOLITE GROUP

$(Ca_{0,5},Na,K)_5Al_5Si_{19}O_{48} \cdot 13\ H_2O$

MONOCLINIC ● ●

Properties: C – colorless, white, yellowish; S – white; L – vitreous to pearly; D – transparent to translucent; DE – 2.1; H – 4-4.5; CL – good; F – uneven to conchoidal; M – prismatic crystals and complex interpenetration twins.
Origin and occurrence: Hydrothermal in cavities in granitic pegmatites, associated with albite, petalite and elbaite, also as pseudo-morphs after petalite. It occurs in San Piero in Campo, Elba, Italy and in the Opal Hill quarry, Riverside, California, USA.

Ferrierite-Ca

ZEOLITE GROUP

$(Ca,Na,K,Mg)_6Al_6Si_{30}O_{72} \cdot 18\ H_2O$

ORTHORHOMBIC ● ●

Properties: C – colorless, white, greenish, pink, brownish; S – white; L – vitreous; D – transparent to translucent; DE – 2.1; H – 3-3.5; CL – good; F – uneven to conchoidal; M – tabular crystals, platy and columnar aggregates, massive.
Origin and occurrence: Hydrothermal in cavities in volcanic rocks, along the cracks of the Alpine-type veins and in volcanic tuffs, associated with calcite and zeolites. Platy aggregates occur in Svojanov, Czech Republic; Albero Bosso and Monastir, Sardinia, Italy; and Kamploops Lake, British Columbia, Canada.

Ferrierite-Ca, 2 mm aggregates, Regéc, Hungary

Dachiardite-Ca, 2 mm xx, Regéc, Hungary

Laumontite

ZEOLITE GROUP

$Ca_4Al_8Si_{16}O_{48} \cdot 18\ H_2O$

MONOCLINIC ● ● ● ●

Properties: C – colorless, white, yellowish, pinkish; S – white; L – vitreous; D – transparent to translucent, weathered almost opaque; DE – 2.4; H – 3-4; CL – good; F – uneven to conchoidal; M – prismatic crystals, fibrous and radial aggregates.
Origin and occurrence: Hydrothermal in cavities in volcanic rocks, along the cracks in the Alpine-type veins, in ore veins and sediments, associated with calcite and other zeolites. Colorless and white acicular crystals up to 38 cm (14 15/16 in) long come from the Pandulena Hill Quarries, India. Crystals up to 150 mm (6 in) long known from Pine Creek, Bishop, California, USA. It was also found in Dillenburg, Germany.

Laumontite, 120 mm, Markovice, Czech Republic

Heulandite-Ca, 8 mm xx, Paraná, Brazil

Heulandite-Ca

ZEOLITE GROUP

$(Ca_{0,5},Sr_{0,5},Na,K)_{4,5}Al_9Si_{27}O_{72} \cdot 24\ H_2O$

MONOCLINIC ● ● ● ●

Properties: C – colorless, white, yellowish, pinkish, red, brown; S – white; L B vitreous to pearly; D – transparent to translucent; DE – 2.1; H – 3.5-4; CL – perfect; F – uneven to conchoidal; M – prismatic to tabular crystals, granular, massive.

Origin and occurrence: Hydrothermal in cavities in volcanic rocks, in the Alpine-type fissures, ore veins and sedimentary rocks, usually associated with calcite and other zeolites. Colorless and white tabular crystals up to 100 mm (4 in) across come from Nasik, India; West Paterson, New Jersey, USA; and Teigarhorn, Iceland. Red heulandite found in Val di Fassa, Italy.

Clinoptilolite-Ca, 2 mm xx, Honcova Hůrka, Czech Republic

Epistilbite, 15 mm x, Nasik, India

Clinoptilolite-Ca

ZEOLITE GROUP

$(Ca_{0,5},Na,K)_{4,5}Al_9Si_{27}O_{72} \cdot 24\ H_2O$

MONOCLINIC ● ● ● ●

Properties: C – colorless, white, yellowish, pinkish, red, greenish; S – white; L – vitreous; D – transparent to translucent; DE – 2.2; H – 3.5-4; CL – perfect; F – uneven to conchoidal; M – platy crystals, massive.
Origin and occurrence: Hydrothermal in volcanic-sedimentary rocks. Large industrial deposits occur in New Zealand, Japan and Australia. Crystals come from Agate Beach, Oregon, USA.
Application: construction, chemical industry and agriculture.

Epistilbite

ZEOLITE GROUP

$(Ca,Na_2)Al_2Si_4O_{12} \cdot 4\ H_2O$

MONOCLINIC ● ● ●

Properties: C – colorless, white, yellowish, pinkish, light brown; S – white; L – vitreous; D – transparent to translucent; DE – 2.2; H – 4; CL – good; F – uneven to conchoidal; M – prismatic crystals, radial aggregates.
Origin and occurrence: Hydrothermal in cavities in volcanic rocks; associated with other zeolites. Clear and white tabular crystals and their twins up to 30 mm (1 3/16 in) across come from Jalgaon, India; also in Teigarhorn, Iceland and Faeroe Islands.

Stilbite-Ca

ZEOLITE GROUP

$(Ca_{0,5},Na,K)_9Al_9Si_{27}O_{72} \cdot 28\ H_2O$

MONOCLINIC ● ● ● ●

Properties: C – colorless, white, yellowish, pinkish, brown; S – white; L – vitreous, locally pearly; D – transparent to translucent; DE – 2.2; H – 3.5-4; CL – good; F – uneven to conchoidal; M – prismatic and tabular crystals, commonly complicated interpenetration twins and sheaf-like aggregates, granular, massive.
Origin and occurrence: Hydrothermal in cavities of volcanic rocks, along Alpine-type fissures, in ore veins, also in sedimentary rocks and hot springs, typically associated with calcite and other zeolites. Colorless and white tabular crystals and their combinations up to 200 mm (7 7/8 in) across come from the Pandulena Hill Quarries, India. Crystals are also known from Teigarhorn, Iceland; Faeroe Islands; West Paterson, New Jersey, USA and elsewhere.

Stilbite-Ca, 83 mm, Nasik, India

Gismondine, 2 mm xx, Tapolca-Diszel, Hungary

Phillipsite-K, 3 mm xx, Limberg, Germany

Gismondine

ZEOLITE GROUP

$CaAl_2Si_2O_8 . 4{,}5\ H_2O$

MONOCLINIC ● ●

Properties: C – colorless, white, bluish, pinkish; S – white; L – vitreous; D – transparent to translucent; DE – 2.3; H – 4.5; CL – good; F – uneven; M – complex twins of crystals, platy aggregates.
Origin and occurrence: Hydrothermal in cavities in volcanic rocks, in hydrothermally altered rocks. It comes from Capo di Bove, Italy; Schiffenberg, Germany; and Dobrná, Czech Republic.

Harmotome, 79 mm, Strontian, UK

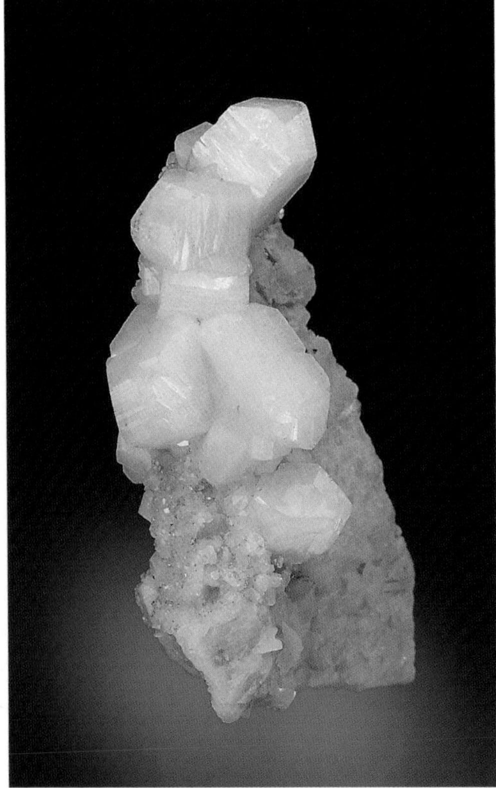

Phillipsite-K

ZEOLITE GROUP

$(K,Na,Ca_{0,5})Al_5Si_{11}O_{32} . 12\ H_2O$

MONOCLINIC ● ● ●

Properties: C – colorless, white, reddish; S – white; L – vitreous; D – transparent to translucent; DE – 2.2; H – 4-4.5; CL – good; F – uneven; M – prismatic crystals and complex interpenetration twins, granular, massive.
Origin and occurrence: Hydrothermal in cavities in volcanic rocks, sedimentary rocks and hot springs; associated with calcite and other zeolites. Colorless and white interpenetration twins up to 20 mm ($^{25}/_{32}$ in) across come from Capo di Bove, Italy; Doughboys, Tasmania, Australia; Sovinec near Litomisce, Czech Republic and elsewhere.

Harmotome

ZEOLITE GROUP

$(Ba_{0,5},Ca_{0,5},K,Na)_5Al_5Si_{11}O_{32} . 12\ H_2O$

MONOCLINIC ● ● ●

Properties: C – colorless, white, gray, reddish, yellow, brown; S – white; L – vitreous; D – transparent to translucent; DE – 2.4; H – 4.5; CL – good; F – uneven; M – prismatic crystals, commonly complex interpenetration twins, granular massive.
Origin and occurrence: Hydrothermal in cavities in volcanic rocks, in the Alpine-type veins, ore veins and in pegmatites. Prismatic crystals and their twins up to 20 mm ($^{25}/_{32}$ in) across are known from Strontian, Scotland, UK. Crystals also come from St. Andreasberg and Oberstein, Germany; Kozákov, Czech Republic; and Kongsberg, Norway.

Goosecreekite

ZEOLITE GROUP

$CaAl_2Si_3O_{10} . 5\ H_2O$

MONOCLINIC ●

Properties: C – white, colorless; S – white; L – vitreous; D – transparent to translucent; DE – 2.2; H – 4.5; CL B perfect; F – uneven to conchoidal; M – prismatic crystals, granular.
Origin and occurrence: Hydrothermal in cavities in volcanic rocks. It is known from Goose Creek, Virginia, USA. Crystals up to 30 mm (1³/₁₆ in) across found in the Pandulena Hill quarries, India.

Goosecreekite, 25 mm xx, Maharashtra, India

Chabasite-Ca

ZEOLITE GROUP

$(Ca_{0,5},Na,K)Al_4Si_8O_{24} . 12\ H_2O$

TRIGONAL ● ● ●

Properties: C – colorless, white, yellowish, pinkish, greenish; S – white; L – vitreous to dull; D – transparent to translucent; DE – 2.2; H – 4-4.5; CL – imperfect; F – uneven to conchoidal; M – rhombohedral crystals and their twins, granular, massive.
Origin and occurrence: Hydrothermal in cavities in volcanic rocks and pegmatites, along Alpine-type fissures and in hot springs, usually associated with calcite and other zeolites. Colorless and white interpenetration twins up to 60 mm (24 in) across, come from Faeroe Islands. Crystals are also known from Řepčice, Czech Republic; Berufjord, Iceland; Panvil, India; Maglovec, Slovakia and elsewhere.

Chabasite-Ca, Passboro, Nova Scotia, Canada

10. Organic compounds

Whewellite

$CaC_2O_4 . H_2O$

MONOCLINIC ● ● ●

Properties: C – colorless, white, grayish; S – colorless; L – vitreous; D – transparent to translucent; DE – 2,2; H – 2,5; CL – good; F – conchoidal; M – prismatic crystals, commonly twinned.
Origin and occurrence: Rare hydrothermal, mainly sedimentary in coal basins, associated with barite, ankerite and other minerals. Hydrothermal crystals up to 70 mm (2¾ in) long, come from U-bearing veins in Príbram, Czech Republic. Similar crystals, up to 70 mm (2¾ in) long, occurred in Cavnic, Romania. Heart-shaped and butterfly twins, up to 100 mm (4 in) found in concretions near Kladno, Czech Republic. Similar specimens are known from Burgk near Dresden, Germany. Interesting flat radial aggregates found along cracks in clays in the vicinity of Most, Czech Republic.

Amber, 40 mm, Baltic Sea, Latvia
Whewellite, 27 mm x, Burgk, Germany

Mellite, 170 mm, Bicske-Csordakút, Hungary

Mellite

$Al_2[C_6(COO)_6] . 18\ H_2O$

TETRAGONAL ● ●

Properties: C – honey-yellow; S – white; L – resinous to vitreous; D – transparent to translucent; DE – 1,7; H – 2; CL – imperfect; F – conchoidal; M – dipyramidal crystals, granular; LU – blue.
Origin and occurrence: Secondary in the cracks in brown coal and lignite. The best specimens with crystals up to 40 mm ($1^{9}/_{16}$ in) in size, come from Csordakút near Tatabánya, Hungary. Crystals, up to 10 mm ($^{3}/_{8}$ in) across, were found in Artern, Germany. Granular aggregates are known from Valchov, Czech Republic.

Evenkite

$C_{24}H_{50}$

MONOCLINIC ●

Properties: C – colorless to light yellow; S – colorless; L – waxy; D – translucent; DE – 0,9; H – 1; CL – good; M – pseudo-hexagonal tabular crystals; R – melts at 50EC (122°F).
Origin and occurrence: Probably secondary. It occurs within geodes near Evenki, Siberia, Russia. Also known from the cracks in altered andesite in Dubník, Slovakia.

Fichtelite

$C_{19}H_{34}$

MONOCLINIC ● ● ●

Properties: C – colorless to yellowish; S – colorless; L – vitreous; PS – transparent to translucent; DE – 1; H – 1; M – thin tabular crystals, scales, crystalline crusts.

Evenkite, 50 mm, Dubník, Slovakia

Origin and occurrence: Secondary, typical mineral of peat-bogs. Its crystals are known from Borkovice near Sobeslav, Czech Republic; also known from Marktredwitz, Germany.

Amber

a mixture of hydrocarbons

AMORPHOUS ● ● ●

Properties: C – honey-yellow, yellow-brown, brown, red-brown, blue, green, black; S – white; L – resinous, dull; D – transparent to translucent, rare opaque; DE – 1,0-1,1; H – 2-2,5; CL – none; F – conchoidal; M – massive irregular or drop-shaped aggregates, nodules and fragments; LU – light blue, yellow.
Origin and occurrence: Amber is a petrified resin from Tertiary and Mesozoic conifers, occurring rarely in sediments. Plant or insect remnants are sometimes found trapped in amber. The most famous localities are located along the southern coast of the Baltic Sea in Poland, Germany, Lithuania, Estonia, Latvia and Russia. The largest masses found weighed up to 10 kg (22 lb). Blue amber is known from the Dominican Republic. Other amber localities are found in Syria, Lebanon, Thailand, Vietnam, Canada and the USA.
Application: as a gemstone.

Fichtelite, 80 mm, Mažice, Czech Republic

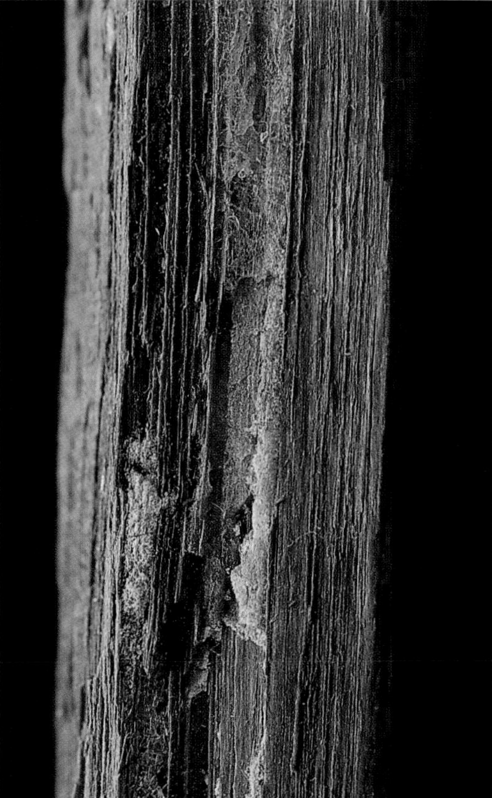

Amber, 95 mm, Madagascar

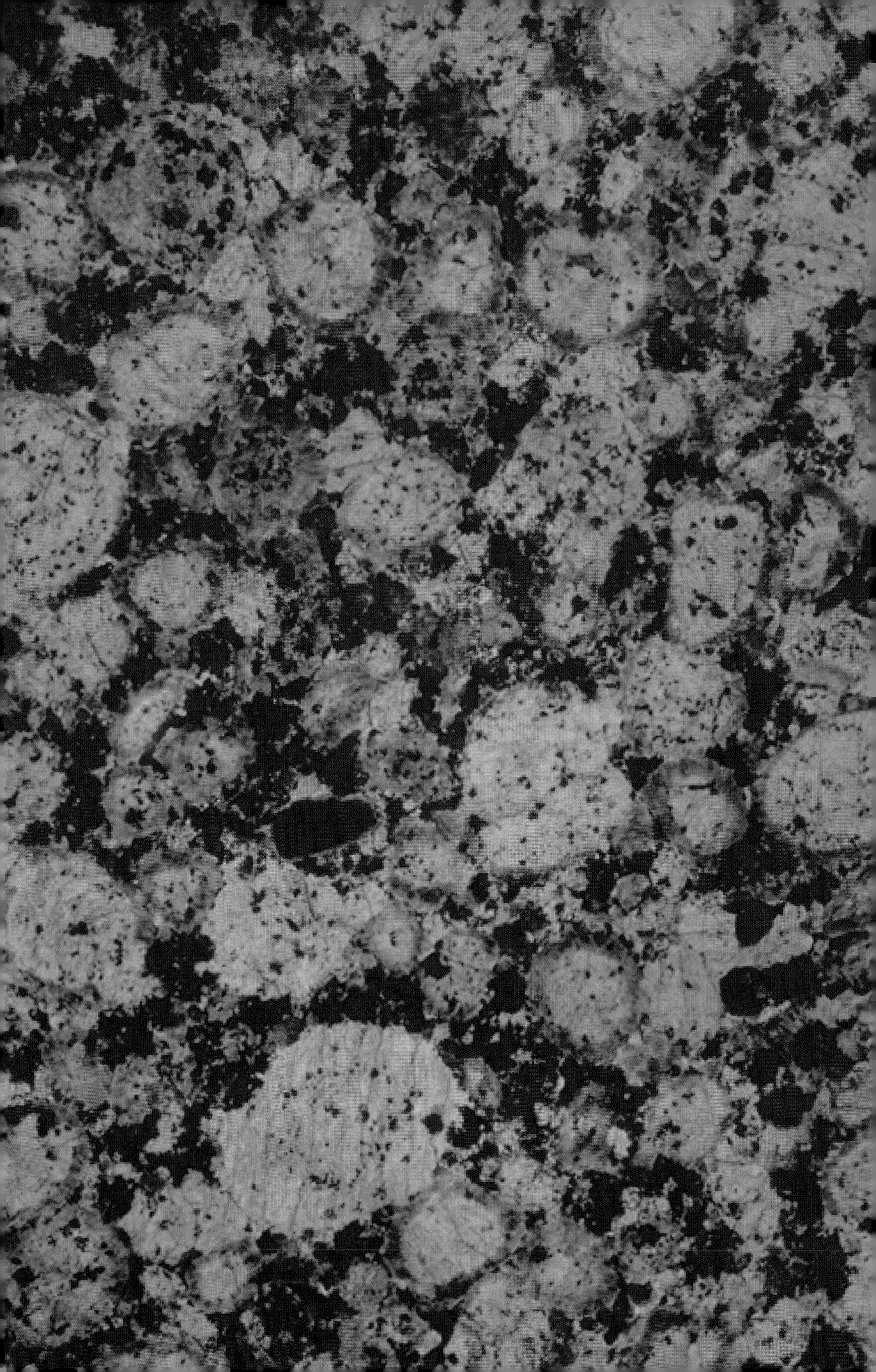

11. Rocks, meteorites and tektites

Rocks consist of minerals and often of fragments of other rocks and organic matter. Since rocks are mixtures, we cannot determine certain physical/ chemical data (chemical formula, crystal system, hardness, etc.). Unlike minerals, some rocks form huge bodies which may cover several thousand km^{2}, so no single localities are shown in this book. Rocks are mainly used for building, in agriculture, and as raw materials for chemicals, ceramics and metals.
Rocks are generally divided into three principal groups according to their origin:
1. **igneous rocks**
2. **sedimentary rocks**
3. **metamorphic rocks**.

Granite, 65 mm, Norway

1. Igneous rocks

Properties: C – light gray, gray, white, brown, black, gray-green, pinkish, red-brown; D – opaque, rare transparent to translucent; L – dull, rare vitreous; DE – varies from 2.6 up to about 3.8; F – uneven, rare conchoidal; M – coarse to fine grained, commonly massive aggregates, consisting of microscopic or larger grains or crystals of various minerals, up to several decameters across, glass is also present rarely. Minerals in rocks usually have characteristic texture features, like graphic granite and others.
Origin and classification: Igneous rocks form at high temperature and commonly at high pressure in solidification of mainly silicate magma of variable composition. The mineral composition reflects a chemical composition of magma. Typical rock-forming minerals are quartz, orthoclase, microcline, plagioclase, biotite, muscovite, amphiboles, pyroxenes, olivine and nepheline. Rocks may solidify at various depths, according to which they are either intrusive, dyke or effusive. They can be grouped according to chemical composition (SiO2 content):
felsic – granite, syenite, pegmatite, rhyolite, obsidian
intermediate – diorite, andesit **basic** – gabbro, basalt
utrabasic – peridotite, varies from 1.0 up to about 2.8; F – uneven, sometimes conchoidal; M – coarse to fine grained and massive aggregates, consisting of grains or crystals, from microscopic to several dm in size, also of rock fragments and organic matter, banded textures are typical and some rocks are fossiliferous.

Orbicular granite, 150 mm, Sweden
Pegmatite, 65 mm, Dolní Bory, Czech Republic

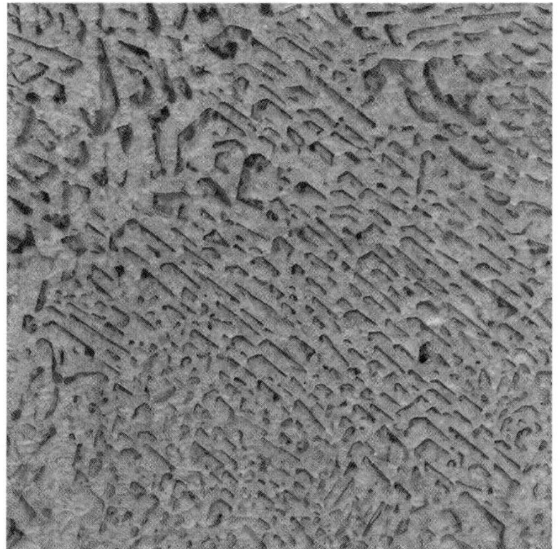

Diabase, 65 mm, Rožany, Czech Republic

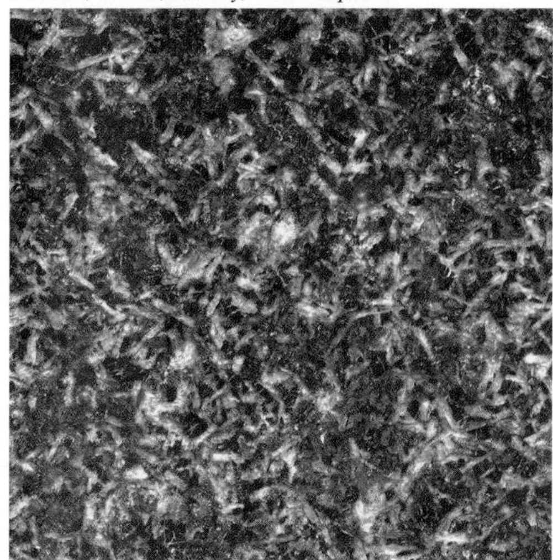

Sandstone, 65 mm, Buffalo Gap, South Dakota, U.S.A.

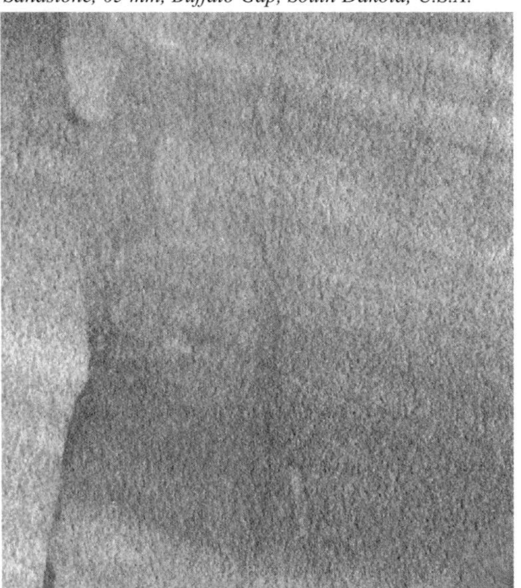

Conglomerate, 65 mm, St.Alban, UK

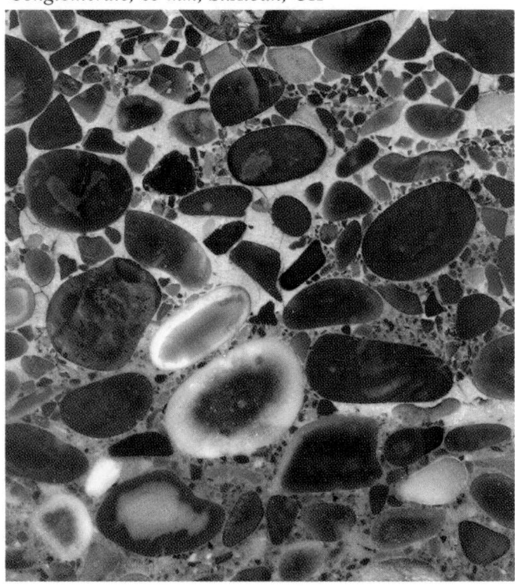

2. Sedimentary rocks

Properties: C – grey, brown, red-brown, black, green-grey, pinkish, white, yellow, often variable even within one rock, banding is typical; D – opaque, rare translucent; L – dull, sometimes vitreous, greasy or earthy; DE -varies from 1.0 up to about 2.8; F - uneven, sometimes conchoidal; M – coarse to fine grained and massive aggregates, consisting of grains or crystals, ranging from microscopic to several dm in size, also of rock fragments and organic matter, banded textures are typical and some rocks are fossiliferous.
Origin and classification: Sedimentary rocks originate under surface temperatures and pressures, as a result of sedimentation of mineral and rock fragments and organic matter of different size through the water and wind activity, or by precipitation from water solutions. Typical rock-forming minerals are quartz, calcite, dolomite, halite, clay minerals and others. According to their origins we can distinguish several groups of sedimentary rocks:
clastic (consisting of rock fragments) – sandstone, conglomerate, quartzite, siltstone.
organic (consisting mainly of organic matter) – limestone, coal.
chemical (originating by precipitation from water solutions) – evaporites, travertine.

Limestone, 65 mm, Italy

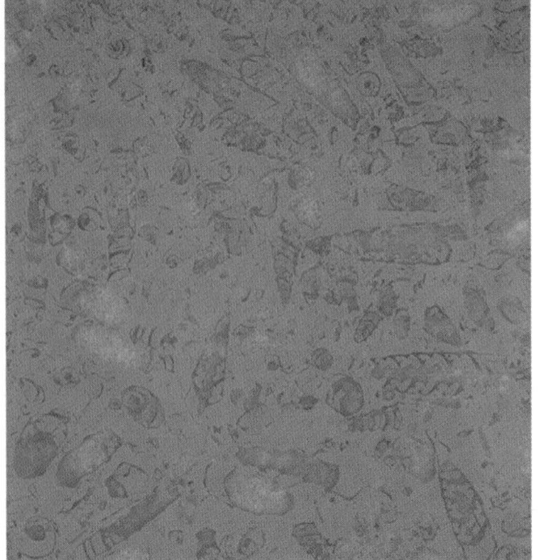

Evaporite, 40 mm, Walkenried, Germany

Serpentinite, 65 mm, Hrubšice, Czech Republic

Gneiss, 65 mm, Doubravčany, Czech Republic

3. Metamorphic rocks

Properties: C – light gray, gray, brown, green, red-brown, black, green-gray, pinkish, white, sometimes variable within one rock, banding is relatively common; D – opaque; L – dull, rare vitreous; DE – varies from 2.5 to 4.8; F – uneven, rare conchoidal; M – coarse to fine-grained, platy, acicular and sometimes massive aggregates, consisting of grains and crystals, ranging from microscopic to several decimeters across. Typical are planar textures and foliation of some minerals.

Origin and classification: Metamorphic rocks originate under higher temperature and pressure during metamorphism of originally igneous or sedimentary rocks. The source of a thermal energy could be magma, then this type is called contact metamorphism, or the thermal source lies in the depth of the earth's crust and effects large areas, then this type is called regional metamorphism. During a process of metamorphism new minerals originate. Typical rock-forming metamorphic minerals are quartz, orthoclase, plagioclases, biotite, muscovite, amphiboles, pyroxenes, calcite, dolomite, sillimanite, kyanite, almandine, staurolite and serpentine.

Regionally metamorphosed rocks: serpentinite, mica schist, gneiss, marble.

Contact metamorphosed rocks: contact chert (porcelanite), skarn.

Marble, 100 mm, Greece

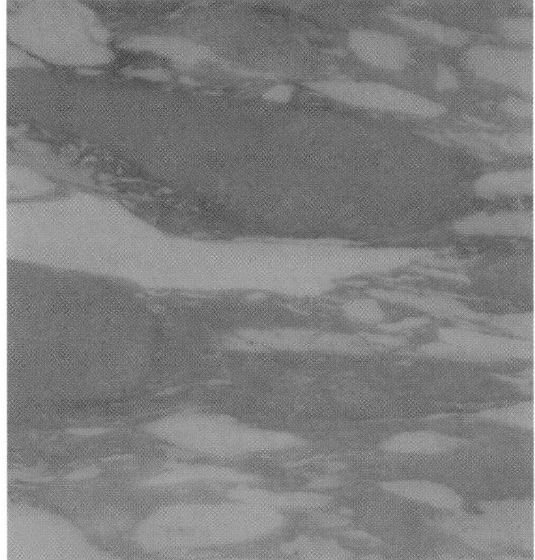

Skarn, 65 mm, Líšná, Czech Republic

Iron meteorite, 50 mm, Sikhote-Alin, Russia

Iron meteorite, 40 mm detail, Gibeon, Namibia

Meteorites

Properties: C - light gray, gray, gray-green, black; D – opaque; L – dull, metallic, rare vitreous; DE – varies from 3.0 to 7.3; F – uneven; M – coarse to fine-grained, sometimes massive aggregates, consisting of irregular grains of different minerals, ranging from microscopic to several cm across.

Origin and classification: Meteorites are igneousrocks formed in space. Most originate in the asteroidbelt between Mars and Jupiter. They consist of various minerals and their chemical composition differs greatly. Typical rock-forming minerals in meteorites are olivine, pyroxenes, plagioclases, Fe and Ni alloys and sulfides, rarely also organic matter. Meteorites fall into four main groups, according to metallic iron and silicate component: iron meteorite, siderolite, chondrite, achondrite.

Occurrence: Meteorite falls are known throughout the world. The largest known iron meteorite, weighing approximately 60 tons, is located near the Hoba farm, Namibia. The largest known chondrite, weighing about 1 ton , fell in 1948 in Norton County, Nebraska, USA. Most meteorites that have been found recently come from large glaciers such as those of Antarctica and from desert in Namibia.

Moldavite, 35 mm, Southern Bohemia, Czech Republic

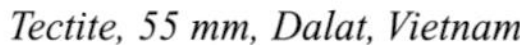

Tectite, 55 mm, Dalat, Vietnam

Tektites

Properties: C – light to dark green, yellow-green, brown-green, brown, green-gray, black; D – transparent, translucent to opaque; L – vitreous; DE – varies from 2.3 to 2.6; H – 6-7; F – conchoidal to uneven; M – massive irregular, drop-shaped or disc-shaped aggregates, irregular fragments, sometimes with typical sculptured surface.
Origin, classification and occurrence: Natural glasses, rich in SiO2. which formed as a result of rapid melting of surface rocks during impacts of large meteorites or comets.
They are classified according to their age and occurrence (following sequence from the oldest to the youngest):
Bediasites and georgianites – the USA, Mexico, Barbados, Cuba
Urengoites – Novyi Urengoi, Russia
Moldavites – southern Bohemia and western Moravia, Czech Republic
Ivorites – Ivory Coast
Irghisites – Zhamanshin, Russia
Indochinites, phillipinites, javanites, bilitonites – southeastern Asia
Australites – Australia
Application: some tektites, mainly moldavites, as gemstones.

Moldavites, 40 mm, Southern Bohemia, Czech Republic

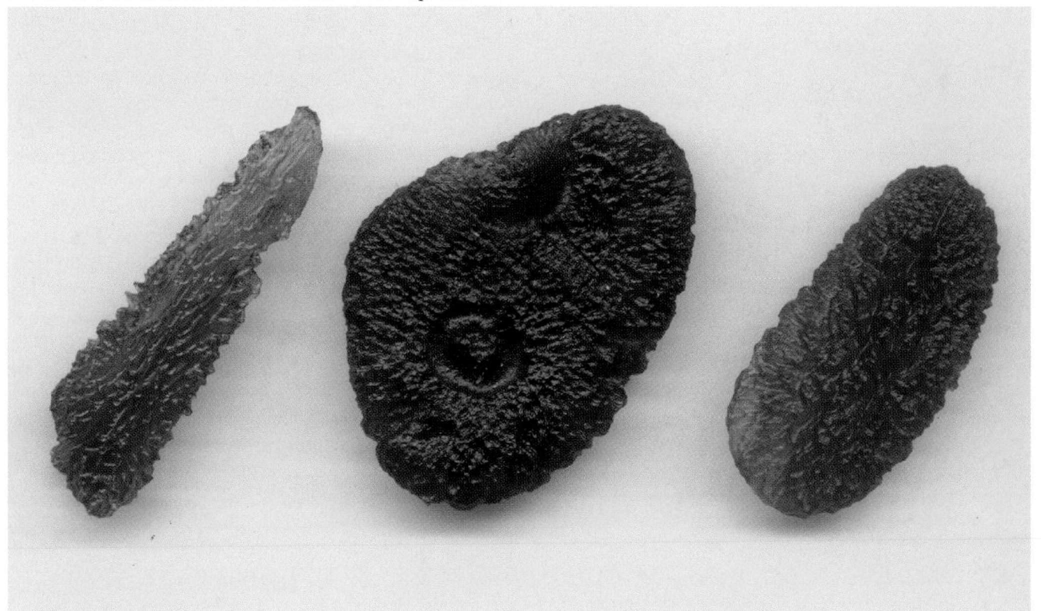

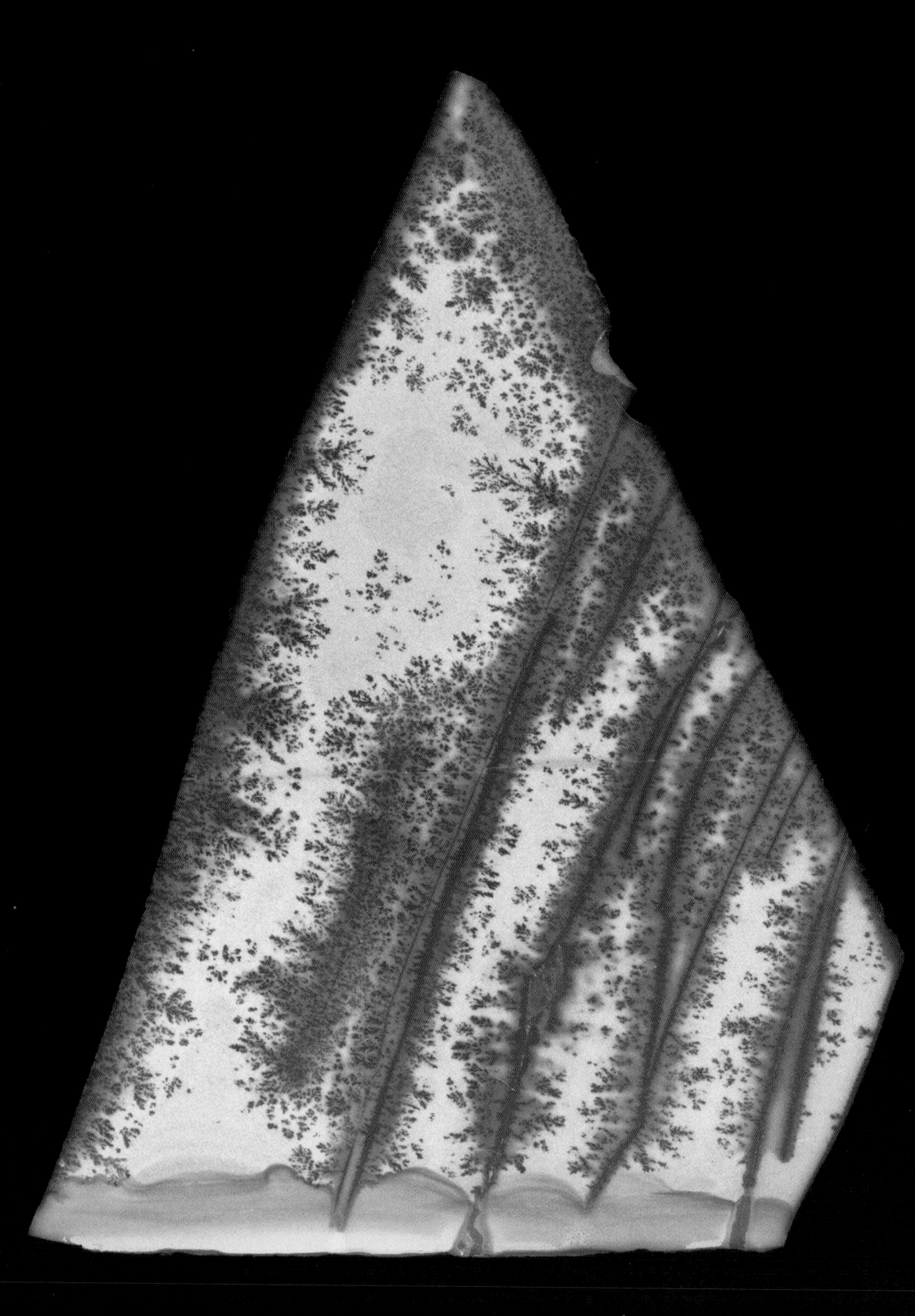

Recommended literature

Anthony J.W. et al. (1990, 1995, 1997):
Handbook of Mineralogy I, II, III,
Mineral Data Publishing, Tucson.

Bernard J.H., Rost R. a kol. (1992):
Encyklopedický přehled minerálů,
Academia, Prague, in Czech.

Clark A.M. (1993):
Hey's Mineral Index, 3rd print,
Chapman & Hall, London.

Duda R., Rejl L. (1986):
Minerals of the World,
Hamlyn, Twickenham.

Fleischer M., Mandarino J.A. (1995):
Glossary of Mineral Species,
Mineralogical Record, Tucson.

Roberts W.L. et al. (1990):
Encyclopedia of Minerals,
Van Nostrand Reinhold, New York.

Strunz H. (1977):
Mineralogische Tabellen, 6th print,
Geest & Portig, Leipzig.

Porcelanite, 85 mm, Komňa, Czech Republic
Agate, 50 mm, Železnice, Czech Republic

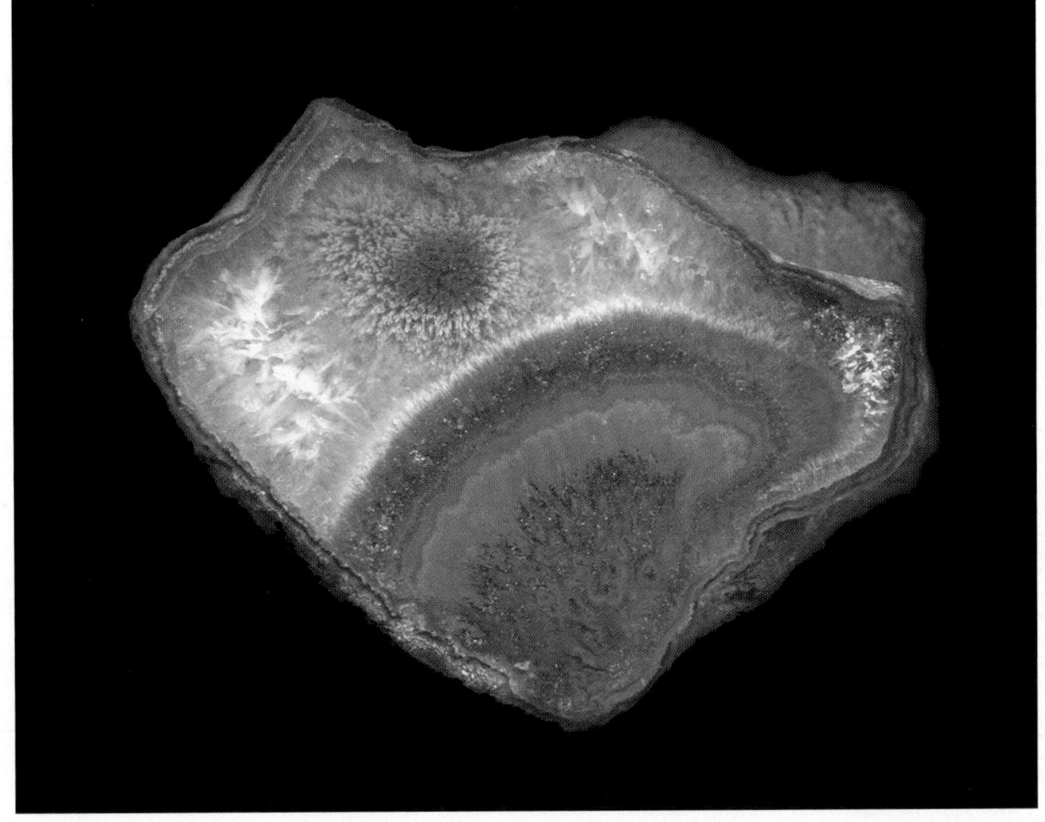

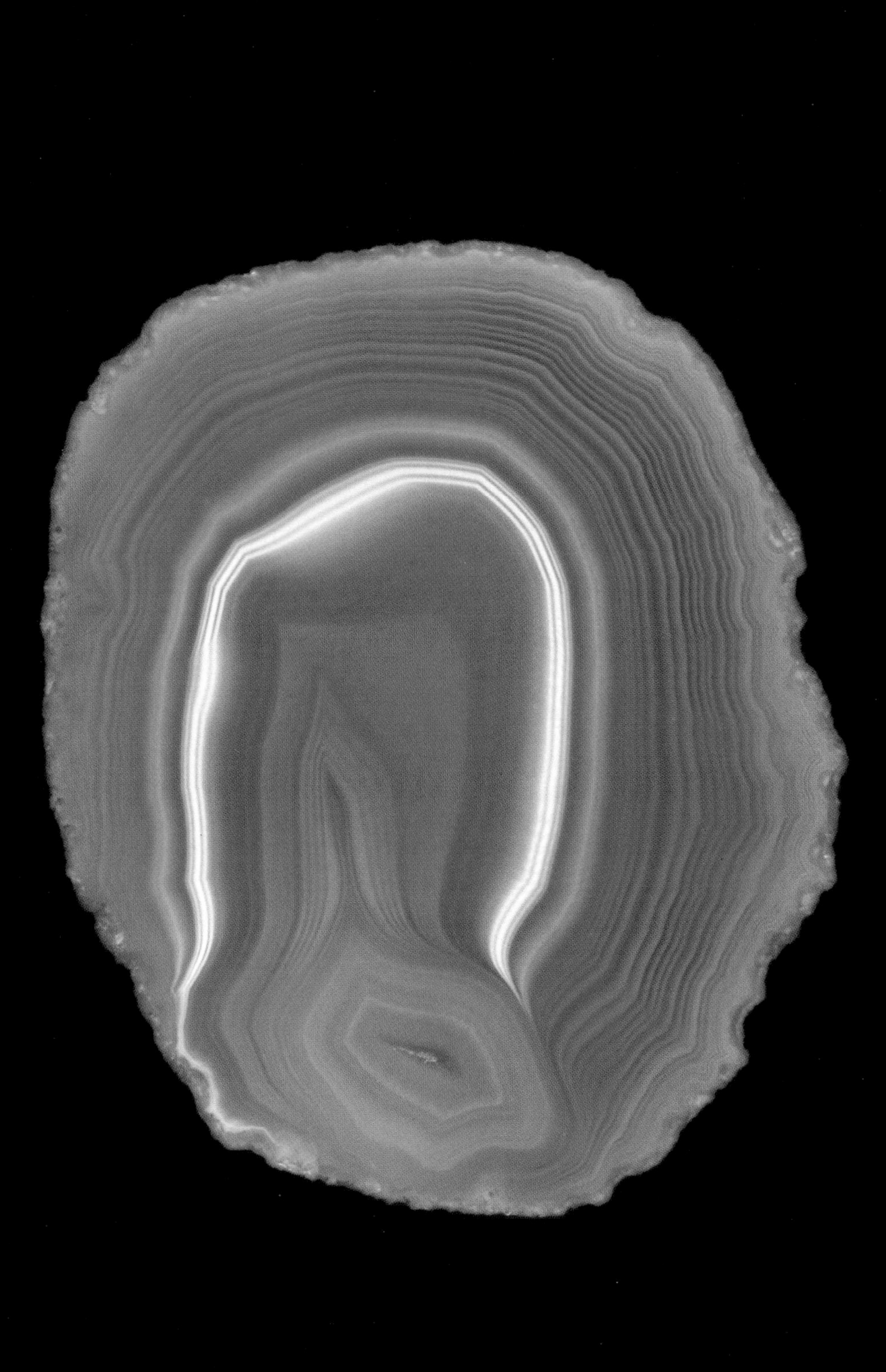

Index